Longman Handbooks in Agriculture

Series editors
C. T. Whittemore
R. J. Thomas
J. H. D. Prescott

Whittemore: *Lactation of the dairy cow*
Whittemore: *Pig production: the scientific and practical principles*
Speedy: *Sheep production: science into practice*

Pig production

The scientific and practical principles

Colin T. Whittemore

Head of Animal Production Advisory and Development, Edinburgh School of Agriculture

Longman   London and New York

Longman Group Limited
Longman House, Burnt Mill,
Harlow, Essex CM20 2JE, England

Published in the United States of America
by Longman Inc., New York

© Longman Group Limited 1980

First published 1980
Reprinted 1982

British Library Cataloguing in Publication Data
Whittemore, Colin Trengove
    Pig production. – (Longman handbooks in
    agriculture).
    1. Swine
    I. Title
    636.4′08   SF395   79–42758

ISBN 0-582-45590-1

Printed in Hong Kong by Astros Printing Ltd

Contents

To the memory of
Frank W. H. Elsley

Preface

Pigs are supreme amongst meat animals, and producing pigs is quite different from other forms of animal production. Pigs are kept intensively; every aspect of their lives under close supervision. Failure to understand their peculiar demands is to fail at pig production. But today, good husbandry alone will not ensure a profit – the industry is one of high science and technology. This must not be confused with the making of rigid rules. Indeed, for animals, science may only become properly objective when choice is allowed. The book, intended for all those with a scientific or practical interest in the pig industry, therefore tries to inform, but not to direct, its readers.

Many friends have been involved in the preparation of the text, but the author would particularly wish to acknowledge the help of Mr A. Gibson, The University of Edinburgh, The East of Scotland College of Agriculture, and his wife. Figures 5.17, 8.11, 8.17, 8.18 and 8.19 are by kind permission of the *Farmers Weekly* (IPC Business Press Ltd.), Fig. 9.1(*a*) by Format, Fig. 9.2 by Commodore Business Machines Ltd., and Figs 8.12, 8.14 and 8.16 by the Scottish Farm Buildings Investigation Unit. All the other photographs were taken by Mr G. Finnie.

C.T.W.
Edinburgh 1979

The plant and the product—pig units and pig meat

1

First, pig production is for providing the pig producer with profit. Even when national statistics present a grey outlook, as they occasionally do, individual producers on top of their business should still maintain a margin; while often the outlook is actually rosy. It is in the nature of the pig industry that margins fluctuate, so an effective and efficient production policy must be maintained. Pig producers must be adept in the application of new technology.

Next, pig production is for providing the human race with edible meat. About one-quarter of earnings in the European Economic Community are spent on food. Spendings on purchases of meat account for 20 to 30 per cent of food spendings and over half the meat bought is pig meat. Pigs rival fowl in the efficiency with which they convert feed into meat for human consumption. It is because pigs convert cereal-based feeds into lean meat so efficiently that, in the UK, pig meat is cheap relative to lamb or beef. As to the range of feedstuffs they will consume, they are the most catholic of animals, and their carcasses go to provide the widest possible range of meat and meat-containing products.

The plant - pig units

Like any manufacturing plant, a pig unit imports raw materials and processes them into products which are exported from the premises (Fig. 1.1). Unlike a factory, some

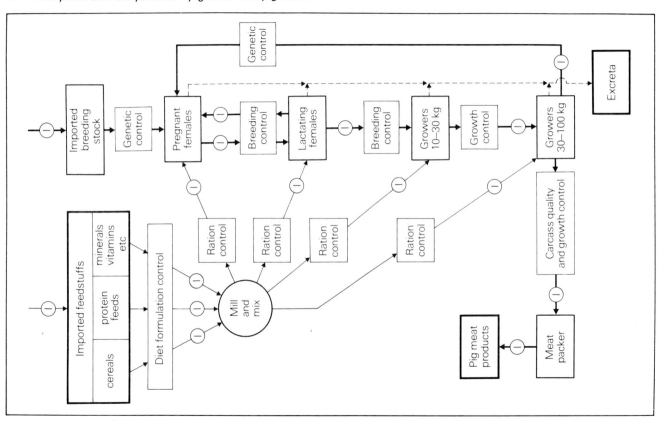

Figure 1.1 Movement of materials within a pig unit. The diagram shows the directions of flow (—→ , food; ——→ , pigs; – – , excreta), control points (☐), and positions for information collection (—①—)

of the raw material is self-generated. Pigs reproduce themselves, although a low level of livestock importation of young breeding animals is usual. The major input is feed. Manpower, buildings, equipment and power are also required, but these comprise only about 20 per cent of total costs. The balance of labour versus capital is tipped well in favour of capital; helped by automatic feeding and excreta removal. At 1980 UK prices, setting up a pig unit for 200 sows could cost the best part of a quarter of a million pounds sterling. The unit would use the labour of two men – say £10 000 per annum – as compared to an interest payment on the capital of around £25 000. The feed bill to supply the necessary feed (1 000 – 1 500 tonnes) would be in the region of £150 000 per annum.

Feedstuffs may come into the unit as individual ingredients such as barley, wheat, maize, soyabean meal and so on, to be mixed together into diets on-site. Or the mixing may be completed off-site and the ready formulated diets purchased in the form of compounded feed. Different types of diet with differing nutrient content are required by different classes of pigs. The various diets are formulated by controlling the rates of flow of ingredients into the milling and mixing facility. Once mixed, diets are put out to the various classes of pigs residing in their respective quarters (Fig. 1.1). Appropriate ration levels are allocated to each pig, or group of pigs, by control of the flow rate of mixed diets into each pig house.

Because feed commands such a large proportion of the total costs of running a pig unit, diet and ration control form the crux of unit management.

Breeding animals move backwards and forwards between houses (Fig. 1.1) as they go through the cycle of pregnancy, parturition, lactation and mating. Age of weaning influences the proportion of total cycle time spent lactating, while mating control dictates the length of the weaning/pregnancy gap and potential litter productivity. As young pigs move on to the growing phase, liveweight gain is practically the sole criterion of adequacy of performance; but in later stages, carcass quality becomes an important, and variable, factor governing the success of the unit.

Movement of feedstuffs (diet formulation) and mixed diets (ration allowance), together with the productivity of the breeding herd and the rate and quality of growth, comprise the main flows of raw materials through the unit. To control and manipulate the production process, information is required. Information about feedstuffs, dietary requirements, ration allowances, reproductive performance, growth responses, carcass quality and pig products (Fig. 1.1). Information gathering is an integral and unavoidable part of logical decision-making activity; without information, decisions degenerate into guesses.

The product – pig meat

Half of all the meat eaten in the world is from pigs. Of the pigmeat produced in Europe, about one-third is cured to provide bacon and ham, another third is sold as fresh pork joints, and the rest goes into sausages, pies, cooked meat and prepared foods.

In comparison with sheep and cattle, pigs are particularly good at producing meat. The UK has 14 million breeding ewes, 5 million breeding cattle (dairy and beef) and rather less than 1 million breeding pigs. These generate, for meat slaughterings, 12 million sheep, 4 million cattle and 15 million pigs. Respective carcass tonnages are 0.2 million (sheep), 1 million (cattle) and 0.8 million (pigs). (About 0.7 million tonnes of poultry meat is also produced – mostly from the broiler industry.) The UK is practically self-sufficient in beef and veal, broiler meat and pork. For mutton and lamb the UK is about 50 per cent self-sufficient and less than 50 per cent when it comes to bacon and ham. In general, the eating habits of Europe reflect little change with beef and veal; while lamb

and bacon consumption is falling and poultry and pork consumption rising.

Pigs may be sold for slaughter at any weight between 50 and 120 kg. At liveweights of less than 50 kg luxury prices must be paid to offset the high rate of fixed costs apportioned to each kilo of carcass yield (Table 1.1), whilst at liveweights above 120 kg feed costs escalate (Fig. 1.2). Butchers cutting pigs for sale as fresh pork require carcasses of a size suitable for family

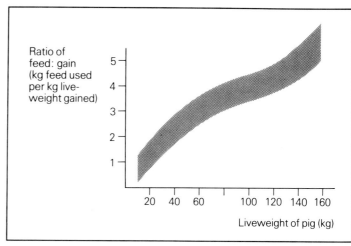

Figure 1.2   As liveweight increases more feed is used for each kilo of weight gain

Table 1.1   Influence of weight at slaughter upon comparative costs of producing a kilo of pig carcass

Liveweight (kg)	Carcass weight (kg)	Fixed costs (p/kg carcass)	Feed costs (p/kg carcass)	Total costs (p/kg carcass)
25	15	170	–	170
50	35	75	21	96
75	55	53	27	80
100	75	40	36	76
125	95	35	42	77
150	115	33	54	87

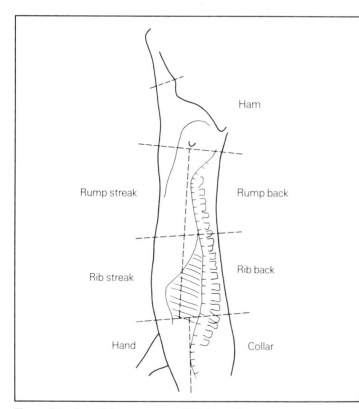

Figure 1.3 One of the many ways of jointing a pig carcass

joints; usually from pigs of 50 to 75 kg liveweight. Pigs sent away to slaughter above 75 kg may either go totally for curing into bacon and ham, or may be cut into parts (for example as in Fig. 1.3), some of which are distributed for fresh sale, some made into sausages, pies and so on, and others cured.

As the pig grows, the intestines and other body organs take up a smaller proportion of the total body mass and the carcass also tends to become fatter, thus the amount going to waste as offals decreases (Fig. 1.4). Generally, carcass weight is about three-quarters of the total liveweight, and the yield of edible meat rather more than half of the liveweight (Fig. 1.5).

The recompense to producers for costs incurred (Table 1.1)

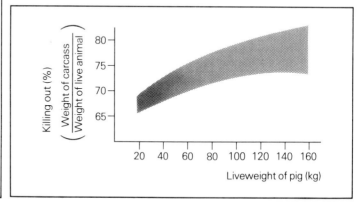

Figure 1.4 Relationship between carcass yield and liveweight

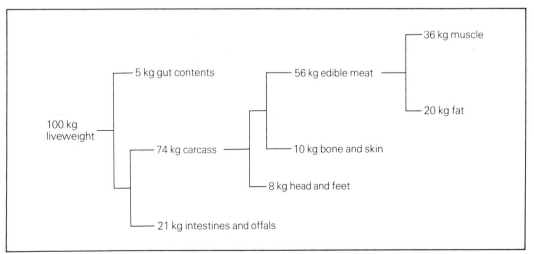

Figure 1.5   Breakdown of 100 kg pig into carcass components

should be balanced to encourage production of pigs over the range of slaughter weights suitable for market requirements. Thus higher prices should be paid for pigs slaughtered below 75 kg liveweight (55 kg deadweight), although often the premium for light pork pigs is inadequate to cover the extra production costs. Where payment is made at a flat rate per kilo of carcass, the benefits are with heavier pigs up to about 120 kg live (95 kg dead). Unless carefully fed, these carcasses can tend to be either fat, or old, and a discerning market may place a lower value upon them. It is not surprising therefore that commercial systems have tended to settle to the production of pigs for slaughter at 70 to 110 kg liveweight (50 to 85 kg deadweight). Despite the importance of the quality aspects of meat production, the major force acting upon profitability is the value of the carcass; the average pig price multiplied by the weight at the sale. Against this should be set the cost of the feed; that is the average feed price multiplied by the quantity of feed used.

Quality

Meat to be sold in developed countries requires above all to be lean. Carcasses are usually graded according to their fatness, and the prices paid adjusted accordingly. Two-thirds of all the fat in the pig's carcass is subcutaneous, so a simple measurement of backfat depth will give a good indication of carcass quality. As pigs get bigger, the depth of fat tends to increase (Fig. 1.6). Usually, fat depth is measured over the eye-muscle halfway down the pig's back with a probe pushed in from the outside, first through the skin and then through the fat layer until the fat/lean interface is reached (Fig. 1.7). The popular place is at P_2 at the midback about 65 mm from the midline. Often more than one measurement is taken, for example, P_1 plus P_3.

Meat packers buying pigs for slaughter are usually interested in other measurements as well. Weight is, of course, vital for efficient handling through a factory system which needs a standard raw material, and a minimum length is usually stipulated when carcasses are needed specifically for curing into bacon sides. There can sometimes be a multiple-hurdle grading system. For this, other measurements are taken of the backfat depth in addition to P_2, for example at loin and shoulder. Pigs achieving top grade on P_2 may thus fall at the loin and shoulder hurdles and be down-graded (Table 1.2). However, when it comes to predicting the amount of fat in a carcass, the addition of loin and shoulder measurements do not seem to improve much upon a P_2 measurement alone.

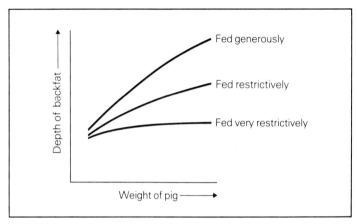

Figure 1.6   As pigs grow, backfat depth increases. The more generous the level of feeding the stronger the relationship

An example of a grading scheme based on carcass weight and P_2 is shown in Table 1.3. As deadweight increases some allowance is made for the natural tendency of pigs to get fatter as they get bigger. The primary concern of the producer is with the maximum fatness governing top grade, and the size of the premium paid for top-grade pigs; on occasion, one or other or both may be unacceptably low.

Females discarded from the breeding herd may be priced at a lower rate. Nevertheless, payment for a used breeding female is usually a significant proportion of the cost of her

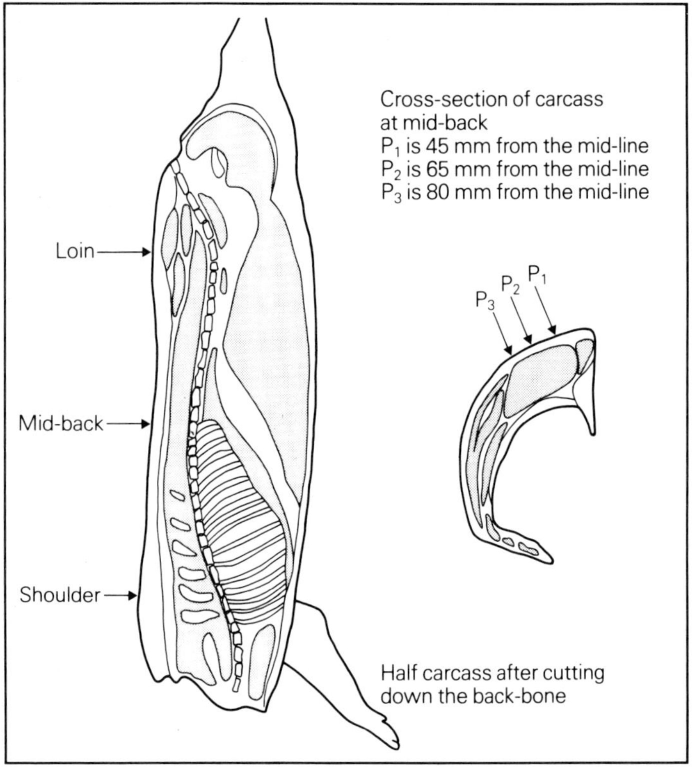
Cross-section of carcass
at mid-back
P1 is 45 mm from the mid-line
P2 is 65 mm from the mid-line
P3 is 80 mm from the mid-line
Loin
Mid-back
Shoulder
P3
P2
P1
Half carcass after cutting
down the back-bone

Figure 1.7 Measurements of backfat depth used for grading pig carcasses.

Table 1.2 Likelihood of a pig being down-graded on fat depth at shoulder or loin after having satisfied the requirement of top grade P_2. In the example, top-grade pigs require to have no more than 18 mm of backfat at P_2, 22 mm at loin and 42 mm at shoulder.

Measurement taken at P_2	Chance of pig being down-graded
14	1 in 20
15	1 in 10
16	1 in 6
17	1 in 4
18	1 in 3

Table 1.3 Example of grading scheme and premiums paid

	Maximum P_2 measurement (mm backfat)			
Weight range (kg deadweight)	Grade 1[a] +4p/kg[c]	Grade 2 +2p/kg	Grade 3	Grade 4[b] –3p/kg
35–45	10	12	15	
45–55	12	15	18	
55–65	15	18	21	
65–75	18	21	24	
75–85	21	24	27	
85–95	24	27	30	
95–105	27	30	33	

[a] Grade 1 pigs may also have a minimum fatness stipulated, for example 12 mm P_2 at 55–75 kg deadweight.
[b] All pigs failing to achieve grade 3 standard.
[c] The premium, or penalty, is added to the going price which will relate to carcass weight and also to prevailing market conditions.

replacement, which casts doubt upon longevity as an asset in the breeding herd. Meat from adult breeding males is not usually acceptable for human consumption due to the coarseness of the meat and the strength of the flavour.

Virgin entire males, and occasionally virgin females and castrated males, may carry an odour, particularly noticeable upon cooking. This led to the view that all males were unacceptable for the production of prime meat. However, modern production techniques, allowing rapid growth, render such ideas defunct. Meat from fast-grown entire mates, reared under appropriate conditions, and slaughtered at no greater than 100 kg liveweight, yield carcasses which are equally acceptable, and often preferable to, those from castrated animals. They invariably carry little fat, sometimes too little. The distribution of joints and the weight of bone may not be quite so favourable, the front-end being a little heavier. It only remains for the pig producer and the pigmeat processor to reach complete agreement on how the benefits of the entire male can best be exploited to the mutual benefit of both.

The abhorrence of meat from entire males in many, particularly developing, countries results from the possibility of unpleasant odour in the carcasses. Any retreat from a policy of castration in such circumstances could harm the pig industry. But there is little doubt that in the advanced industrial climate of British agriculture, considerable positive benefits have accrued to all sectors of pig production from a swing towards the use of entire males for meat. In particular, entire males convert feed into liveweight gain much more effectively than castrates, and are more economical to produce.

Excessively lean pigs – pigs with little backfat – tend to suffer from low fat quality. The symptoms are fat which is soft, which tends to split and which divides away from the muscle. The cure is to ensure that the animal eats more food.

The symptoms of reduction in lean meat quality are usually muscle which is either pale, soft and exudative (watery) ('PSE') or dark, firm and dry ('DFD'). Both these phenomena make the meat unattractive in appearance and less palatable to eat. However, pale meat which is not wet is acceptable, and may even be preferred by some purchasers of fresh pork joints. Neither PSE or DFD is especially critical in the UK, although PSE appears to be associated with specific strains of pig. Both can be reduced, or even eliminated, by care for the welfare of the live pig and the warm carcass. Most problems arise from unacceptable treatment of the pigs during loading and dispatch from the farm, in the course of transportation, on arrival at the meat factory, or immediately prior to slaughter. Reduction in quality of meat also follows from mishandling carcasses after slaughter, and during evisceration, jointing and chilling.

Reproduction

2

The cost of providing food and facilities to a breeding female is almost totally independent of her productivity. Financial success is therefore largely dependent upon the number of piglets produced annually from each female.

Events in the life of a breeding pig are laid out in Fig. 2.1. The timing of some of these events is relatively constant while others are more variable (Fig. 2.2). Of necessity, pregnancy requires the female to be post-pubertal, which, if she is white, is unlikely until after 150 days of age and 80 kilo liveweight (Fig. 2.2(1)). Pregnancy rarely lasts for less than 112 or for more than 117 days (Fig. 2.2(3)); and as pigs cannot conceive while they are lactating, pregnancy and lactation cannot run concurrently. After weaning, the shortest interval until the beginning of the next pregnancy is three or four days (Fig. 2.2(5)). Much more variable, and largely at the discretion of the producer, are the weight or age at which the female is mated for the first time (Fig. 2.2(2)), the length of the lactation (Fig. 2.2(4)) and the length of the interval between the end of lactation and the beginning of the next pregnancy (Fig. 2.2(6)).

The female which is neither lactating nor pregnant will normally be intent on getting pregnant. To this end her hormones will be going through the oestrus cycle (Fig. 2.3). This cycle begins at puberty and continues to conception. It starts again at weaning, which triggers development of the ova to be released some five days later. The female prepares for fertilization by behavioural changes associated with the one-

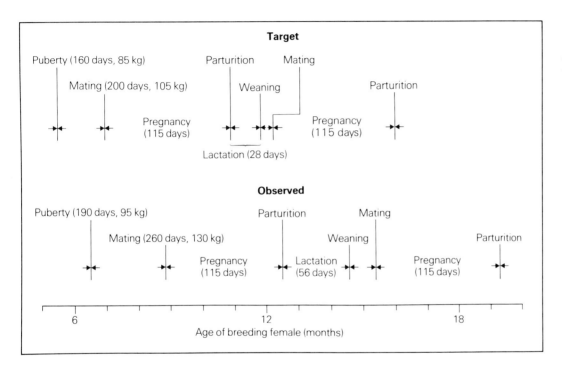

Figure 2.1 Events in the life of breeding female pig. The lower half of the diagram gives some values observed in practice, while the upper half of the diagram gives recommended, and achievable, targets. The commercial average lies somewhere between the two

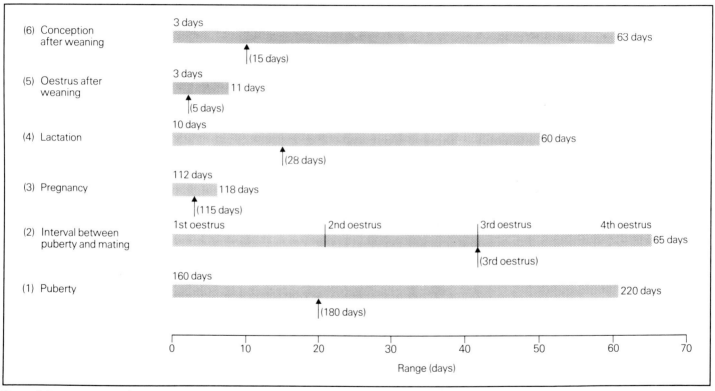

Figure 2.2 Aspects of reproductive performance. Time-scale ranges found in practice. ↑ marks the average position

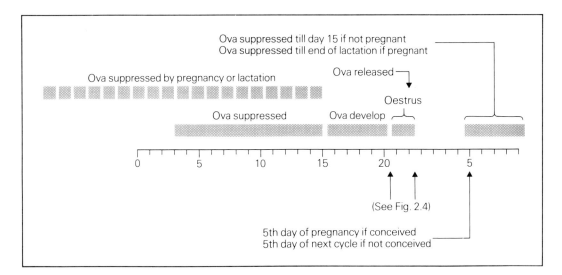

Figure 2.3 Twenty-one day oestrus cycle of pig

Ova suppressed till day 15 if not pregnant
Ova suppressed till end of lactation if pregnant

Ova suppressed by pregnancy or lactation

Ova released

Ova suppressed

Ova develop

Oestrus

0   5   10   15   20   5

(See Fig. 2.4)

5th day of pregnancy if conceived
5th day of next cycle if not conceived

to two-day oestrus period during which she will accept the male for mating (Fig. 2.4). In the natural course of events the male will be allowed to mate before the ova are shed. Indeed the act may help their release, and male sperm need to mature in the female tract before meeting the ova moving down. Given the opportunity, the pair will mate frequently: for as long as the female will allow and as long as the male is able. The *final* act is likely to occur toward the end of oestrus, that is after ova release. Under domestication, such a liberal attitude toward mating results in wasteful use of the male, and it is possible for other females simultaneously in oestrus to remain unmated.

Although the distinction of contributing to the next generation goes to only one sperm, many are required to break down the resistance of the ova to penetration, and there are 20 or more ova to be fertilized. A fertile union in pig production

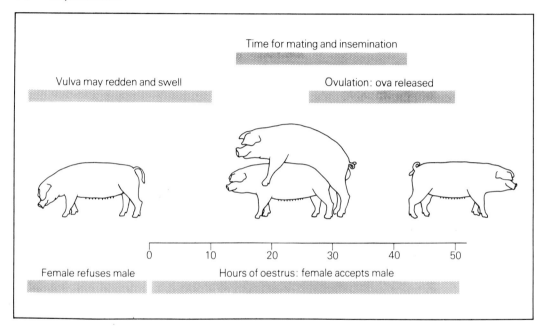

Figure 2.4 Oestrus (heat)

Time for mating and insemination

Vulva may redden and swell

Ovulation: ova released

0   10   20   30   40   50

Female refuses male

Hours of oestrus: female accepts male

terms – where not just *any* pregnancy will do – requires the presence of a high concentration of active sperms at the right place and the right time. Oestrus may last for as short as a few hours and as long as a few days, and it is never possible to know exactly when the best time for service is, as this appears to be about two-thirds through the oestrus period. Mating at the right time can therefore only be ensured by encouraging frequent mating at regular intervals of 12 hours or so during the course of the oestrus period. Mating twice (or artifically inseminating twice) increases the number of young born by about one per litter, and more frequent mating (say, three times) will evoke further advantage. The most efficient way of determining whether or not a female is in oestrus is to put a male first in front of her, and then behind. The next best way is to press down on her back with both hands and to sit astride the animal. If she stands firm under human weight, it is even more likely that she will do the same for a male pig. Artificial insemination is well justified on grounds of breed improvement and under these circumstances a reduction in conception rate of about 10 per cent is not so important.

Puberty

The onset of puberty in a female normally occurs at around 180 days of age and 90 kilo liveweight. It is marked by a swelling and reddening of the vulva – which is particularly symptomatic of oestrus in the young female – and a willingness to stand and support the male during mating. It is not usual to allow mating at first oestrus as the resulting litter may be of reduced size (Fig. 2.5), and body condition could be depredated by the need to support a pregnancy and a lactation while the animal is still actively growing. The drawbacks to early mating can be countered by special care; and some pig producers work on the basis that *any* animal showing oestrus should be mated. Certainly, to wait until the second or third oestrus before mating is to imply that the first oestrus *seen* was

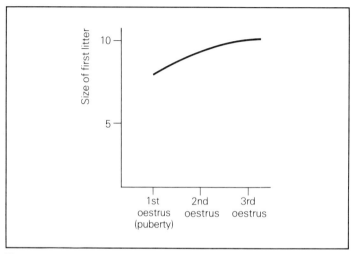

Figure 2.5   Influence of time of first mating on litter size

in fact the first oestrus to occur. Figure 2.2 shows how the range for puberty is such that at any given age a pig might be at first or third oestrus. The 21-day period of the oestrus cycle means that at the next oestrus the pig will be some 10 kg heavier and will have eaten another 50 kg of feed; sufficient incentive for a general rule to the effect that any young female showing oestrus and weighing more than 105 kilo should be mated.

There is some benefit in terms of saved feed from

encouraging puberty at an earlier age, particularly for those potential breeding females which have been grown fast in fattening pens. Puberty can be brought forward by ensuring there are no entire males in the vicinity of females while they are growing, and then allowing, at the chosen time, sudden flash exposure of the male. The sight, sound and smell of the male helps to bring immature females to puberty 10 to 20 days earlier. Disruptions also seem to be effective in this respect; shifting the pigs about, mixing up groups and changing their quarters.

It is generally accepted that the first litter will be smaller than those from experienced females (Fig. 2.6), best performance being reached by the fourth litter, although the effect varies greatly between individual breeding units.

Pregnancy

The fertilized ova travel down the female tract where, in the protected uterine environment, they begin to form into embryos. By about the 25th day of pregnancy these will have implanted into the receptive wall of the uterus. Nourishment passes from the mother to the developing foetus via the placenta, and by the end of pregnancy the conjoining of the microscopic ovum and sperm cells have resulted in a 1.0–1.5 kg baby pig.

The time spent in the uterus is about half the total life of a growing pig destined for meat production, and it is not without its hazards. Most of the growth of the unborn young is during the last 30 days of pregnancy (Fig. 2.7), while most of the losses occur before the foetal stage (Fig. 2.8). It is quite

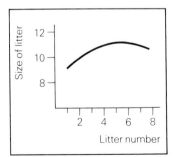

Figure 2.6   Influence of maternal age on litter size

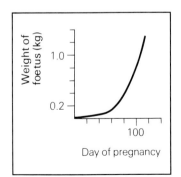

Figure 2.7   Foetal growth

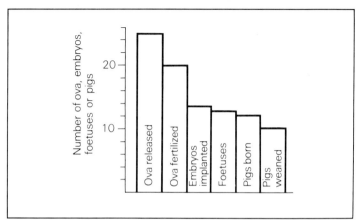

Figure 2.8 Losses during pregnancy. 20 per cent lost by failure to fertilize, 33 per cent lost by failure to implant, 11 per cent of foetuses lost in the uterus, 17 per cent of pigs born lost before weaning

common for 20 to 25 ova to be released from the ovaries during oestrus; of these only 15 to 20 will be fertilized, depending upon the number of sperm in the upper female tract at the time. The next hurdle is for the fertilized ova to implant securely to the uterine wall where they can be nurtured; the failure rate here can be 20 per cent to 40 per cent, depending on the receptivity of the uterus.

While this phase of the pig's life is only poorly understood it does seem that implantation is influenced by the mother's environment, particularly her feeding, housing and welfare. The physiological state of the uterus is also critical to embryo implantation, and this is partly dependent upon the interval between parturition and conception. Once safely implanted, subsequent losses of potential pigs are relatively small, unless some catastrophic event befalls the mother. Numbered among such events would be severe undernutrition, stress or disease, or the ingestion of feeds containing toxic substances. The consequences are often seen as low numbers born, and as dead and partly decomposed foetuses at birth.

Successful management in the breeding herd is usually equated with low mortality amongst baby pigs, and with a high number of litters produced per breeding female per year. This has led to considerable attention being given to the survival of baby pigs and to earlier weaning. Both these considerations can be less significant than the initial litter size (Table 2.1). Nevertheless, there is a diminishing response to improvements in litter size, and most benefit would be gained in herds who average less than 12 live pigs born. Increasing the ovulation rate is unlikely to have much effect in other than maiden females (which may shed fewer ova and therefore benefit from stimulation of ova production by generous feeding for the fortnight before mating). Care of the female around mating to ensure a high embryo implantation rate is clearly crucial to subsequent litter size (Fig. 2.8); the adult female warrants ample feed after weaning and should come

Table 2.1   Factors influencing the number of young pigs weaned annually per breeding female (over an average of 19.5)

Factor	Action	Result (increase in young pigs weaned annually per breeding female)
Post-weaning mortality	Reduce by half (from 20% to 10%)	+ 2.5
Lactation length	Shorten by half (from 4 to 2 weeks)	+ 2
Numbers born	Increase by quarter (from 11 to 14)	+ 6

into oestrus in rising body condition. Unfortunately the definition of 'care' in this context can go no further than platitudes, and there is no doubt that this matter urgently requires attention.

Parturition

The young are born invariably within two days of the 115th day after conception, even allowing for a day's error from the assumption that conception occurred at first mating. The female may, but does not always, show signs of restlessness as term approaches, and if given bedding material will attempt to build a nest. Milk may be available at the mammae from two days to two hours prior to the parturition. The signs are not always particularly apparent, and the best guide is the calendar.

At birth, pigs weigh between 0.75 and 1.75 kg, but most are between 1.0 and 1.5 kg. Pigs weighing less than 1 kg have a lower chance of survival. Litters of more than 13 tend to contain a higher proportion of lighter pigs, due perhaps to a limited uterine capacity. Large litters also tend to have a greater number born dead.

Post-natal mortality may amount to 10 to 20 per cent of those born and often seems to be due to crushing under the mother. The mother's apparent disregard for her young should be tempered by the knowledge that many piglets which end up crushed were already weak at birth. Some form of restraint imposed upon the mother may reduce, but will not prevent, the crushing of baby pigs. Unconfined breeding females are not especially liable to infanticide, unless the quarters are cramped. Close confinement such that the mother cannot turn around or move more than 2 to 3 feet forwards and backwards is convenient for handling and housing, and may reduce some losses. But mortality statistics do not confirm that close confinement of the breeding female has, of itself, cured the problem of deaths in baby pigs.

A large proportion of mortalities amongst baby pigs, whether crushed or not, can be attributed to failure to suck within the first few hours of life, and their succumbing first to cold (to which they are very susceptible), next to rapid starvation, and then to one of the range of pathogenic organisms to which the youngster is unavoidably exposed. When first born, the pig has no fat and little hair to keep itself warm, only a few grams of liver glycogen as food reserve, and little disease immunity. It is vital for the newborn baby pig to lose no time in sucking from his mother.

Lactation

Milk is available at the mammary glands from the beginning of parturition, and as each pig is born it will make its way directly to the mammae, find a nipple and suck. The pattern of lactation in the pig and the interrelationships between mother and young are particularly fascinating, and a full account is given within the companion volume in this series entitled *Lactation of the Dairy Cow*. The first milk is not merely the sole source of life-sustaining nutrients to the youngster, but it also contains a high level of antibodies. These are able, in the first few hours of life, to escape digestion and pass intact through the intestine wall to give a degree of disease immunity against the numerous pathogens which challenge pigs in the first weeks of life. Another type of antibody is continuously secreted in the milk throughout lactation. This one is not absorbed, but acts in the intestines of the baby pig against bacterial infections of the gut.

The nutritional value of milk is high, containing 8 per cent fat, 5 per cent milk sugar, and 6 per cent protein. It has an energy value of 5.5 MJ/kg and a total solids content of 20 g/100 g fresh milk. The yield of milk varies greatly between females and also depends upon both the stage of lactation

(reaching a peak three to four weeks after parturition (Fig. 2.9)) and the number of pigs sucking (Table 2.2). The first lactation may yield less (Table 2.3).

A litter of 10 pigs, sucking hourly, will draw an average of 7 kg of milk daily from the sow which is equivalent to 0.14 kg

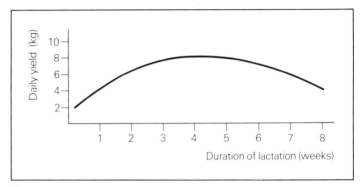

Figure 2.9   Milk yield in relation to age of piglets

Table 2.2   Milk yield in relation to number of sucking young

Number of sucking young	Milk yield (kg/day)	Milk intake (kg/suckler/day)
6	6	1.0
8	7	0.9
10	8	0.8
12	9	0.75

milk solids per sucking pig daily. The growth of sucking pigs is depicted in Fig. 2.10 and the percentage composition of the young pig's body in Fig. 2.11.

Table 2.3   Increase in yield with lactation number

Lactation number	Average milk yield (kg/day)
1	5–6
2	7–8
3	7–8
4	7–8
5	6–7

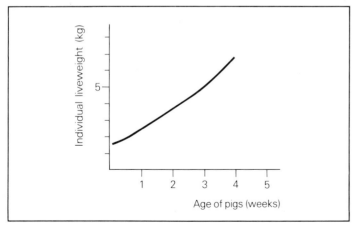

Figure 2.10   Growth of sucking pigs

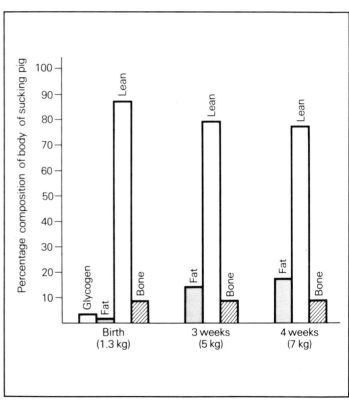

Figure 2.11   Composition of body of sucking pigs

Weaning

Failing the interference of man, mother and young will begin to part company some time subsequent to 12 weeks after parturition. Because pigs cannot rebreed whilst they are lactating, production efficiency necessitates weaning earlier than this and young pigs can be weaned at any age, although usually between three and five weeks. Weaning age depends much upon the degree of sophistication of diet, housing, disease control and management available. Removal at birth or before (by hysterectomy) is used to produce foundation stock for 'specific pathogen-free' 'minimal disease' herds, which have to be guaranteed free of particular pathogenic organisms which pigs would normally pick up as soon as they were exposed to the natural environment (for example, the organisms associated with chronic lung disorders).

Weaning is very much less traumatic for the mother than for the young; the lactating pig has little problem in re-absorbing secreted milk. When the young are removed, gland reversion will begin almost immediately and be essentially complete three days after weaning. This is not altogether surprising as in the first half-day the mammary tissue will have registered as many as 12 missed feeds. There is no necessity for other than instantaneously abrupt weaning, nor any need to withdraw either feed or water from the mother at this time. More likely to be a problem is the condition of the reproductive tract after early weaning. The organs require reconditioning after going through the efforts of nurturing and then expelling a litter. Lactation aids this process, and in any case it is unlikely to be complete until about three weeks after parturition. As the lactation is shortened to be progressively less than three weeks the uterus is correspondingly less receptive to embryo implantation, and there is an increased probability of delayed oestrus and reduced litter size. Although some producers may make a success of weaning at earlier than three weeks for a limited period of time, only a few exceptional units can do so over a number of years. As a general observation, it is difficult to find evidence of any real improvement in number of young pigs produced per breeding female per year from weaning at much less than 25 days of age or 5 kg liveweight.

Weaning to mating

Between weaning and the start of the next pregnancy, the female breeding pig is totally unproductive. Fortunately, the period is usually short – about three to five days – and conception rates in pigs are usually high at 80 to 85 per cent. A few animals may show a delayed oestrus in the second week after weaning, while others may miss the post-weaning oestrus altogether and come into season a few weeks later (Table 2.4). Herd averages for 'empty days' between weaning and conception (including returns, total failures and maiden

Table 2.4 Target conception rates for a pig herd

	Number of pigs		Calculation of herd average	Number of pigs (of 100 weaned)	Days from weaning to conception
	Positive	Negative			
Weaned	100		Conceived at post-weaning oestrus	72	5
Post-weaning oestrus	90	10	Conceived at '21-day' oestrus	21	24
Mated post-weaning	81	9	Mated at '42-day' oestrus	2	45
Conceived and pregnant 21 days later	72	9	Remainder sold at 50th day post-weaning	5	50
		28	Average days from weaning to conception		12
Total not pregnant	28				
Mated at '21-day' oestrus	24	4			
Conceived and pregnant 21 days later	21	3			
		7			
Total not pregnant	7				
Mated at '42-day' oestrus	2				
		5 → Sold off			

replacements) can vary from 12 to 40 days. This difference is worth three to four pigs per breeding female per year. In some units, pigs failing to show signs of oestrus within eight days of weaning are given a hormone injection. The efficacy of this treatment is still debated.

The causes of low conception rates and poor litter sizes, when they occur, are not at all clear and the condition of the tract is only likely to explain *some* reproductive failures in early weaned pigs, and none of the problems of getting breeding females back into pig three weeks or more after parturition. Many insidious breeding problems can be due to breeding females reaching the end of lactation in poor condition and, despite ample feeding in the weaning to mating period, being unable to recover themselves in just one week.

Sometimes breeding herds are subject to storms of infertility. Two agents have been implicated: viruses and fungi. SMEDI viruses are associated with still-births, mummified and aborted piglets, embryo death, and infertility. These therefore will cause problems at every stage of pregnancy. Fungal infections show similar symptoms, and are often associated with improperly stored feed ingredients and a batch of contaminated diet.

Oestrus is encouraged in experienced breeding females by allowing plenty of activity space and interaction; an example would be making up groups of four to eight females and housing them together in a straw yard. Straw yards are not always convenient or available, however, and females awaiting mating may need to be restricted in tethers or stalls. In this case a higher level of management and supervision is needed to achieve satisfactory conception rates. Males should be in close proximity to the females, not merely for ease of arranging mating, but also for the positive effect their presence has upon females. After having been mated two or three times during oestrus there is an 80 per cent or better chance that the female will be pregnant, but no guarantee as to the number of pigs that may be produced 115 days later. To identify the other 20 per cent of females who have not conceived, a rigorous check should be made by an active male between the 18th and 25th days following. Ultrasonic pregnancy detectors used 30 to 40 days after mating, and/or vaginal biopsy techniques, can be useful management aids, but are not infallible. In any event, all females assumed to be pregnant should be closely observed for signs of return to oestrus until such time as they unequivocally demonstrate the fact that they were pregnant by producing young.

Use of males

The male is needed to provide sperm in sufficient quantity and with sufficient frequency to ensure the maximum possible rate of ova fertilization. His presence is also instrumental in bringing females into oestrus in the first place, so that he can mate. To fulfil his functions he must be positioned near at hand to the pens in which the females are held between weaning and confirmed pregnancy; confirmation of pregnancy itself requires a male's physical presence.

The frequency of voluntary mating varies widely amongst males, as does the rate and quality of sperm production. Overwork is the most frequent cause of loss of libido and reduction in sperm concentration and activity.

Males reach puberty at around 150 days and 100 kilo liveweight, and can start work gently (one mating weekly) at about seven months of age, building up to full use over the next two or three months. Practical guidelines to full work have varied from two matings daily to four weekly. Reduction in libido is simple enough to detect, but reduction in fertility following apparently successful mating is harder to spot. A reasonable expectation of work from a male might be 200 to 300 matings yearly; or two females weekly, each of which might be mated two to three times. A male should last for an active working life of two or three years.

The number of males required in a breeding herd depends not so much on the average usage rate, but maximum usage rate. On the basis that it is quite probable for 10 females from a breeding herd of 100 to be in oestrus in the same week, then with one male taking care of two females, five males would be needed for 100 breeding females. These can be distributed through the unit to be available: (i) to virgin females awaiting first mating; (ii) to experienced females immediately post-weaning; and (iii) to females which have not conceived and

are likely to return to oestrus at some later date.

Artificial insemination in pigs is a simple do-it-yourself technique, using fresh semen which has about a three-day refrigerator life and which may be despatched via the postal services. As both conception rate and litter size are about 10 to 20 per cent lower than for natural mating, AI is primarily used for purposes of introducing into nucleus herds specific characteristics possessed by animals of the highest genetic merit standing at AI centres.

Increasing productivity

On most units between 12 and 24 pigs are produced annually from each breeding female. The UK average is around 17 to 18, and has not altered greatly over recent years. As the female eats food and takes up space and labour regardless of her productivity, it costs almost as much to rear 12 young pigs as it does to rear 24. The major factors influencing productivity of breeding females are shown in Fig. 2.12.

Mortality amongst new-born pigs, at or soon after birth, represents a major loss. Deaths at birth comprise about 5 to 20 per cent of pigs born, with a further 5 to 20 per cent dying after birth but within the first week of life. Clearly, there are significant benefits to be derived from keeping mortalities to the lower end of the range.

There are three approaches towards reducing the

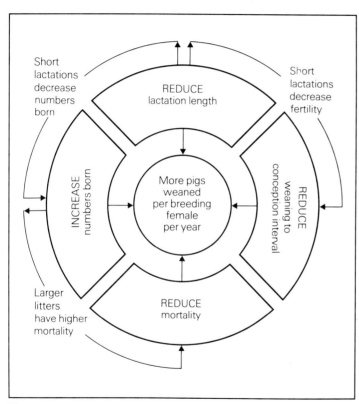

Short
lactations
decrease
numbers
born

REDUCE
lactation length

Short
lactations
decrease
fertility

INCREASE
numbers born

More pigs
weaned
per breeding
female
per year

REDUCE
weaning to
conception interval

Larger
litters
have higher
mortality

REDUCE
mortality

Figure 2.12 Increasing productivity in the breeding herd

predilection that little pigs have for suicide: first, the provision of a warm, acceptable and hygienic environment in the maternity ward to reduce pathogenic organisms; next, adequate care and nutrition of the mother to encourage high individual pig birth weight; and last, a large number of young pigs born would ensure that, despite unavoidable loss, sufficient young nevertheless remain alive.

An increase in the number of pigs born alive in each litter is not easy to attain. It would appear that feeding during pregnancy has less effect on numbers born than on birth weight. There are breed and strain differences in prolificacy – the Large White and White crosses being difficult to surpass. However, between-farm differences are by far the more substantial. It is attention to the details of good husbandry throughout the breeding cycle which ensures minimum losses at the various stages depicted in Fig. 2.8. There appears to be an antagonistic relationship between the number of pigs born in a litter and the duration of the mother's previous lactation if it is less than three weeks. With shorter lactations, gains made from increasing the frequency of parturition may be offset by lower numbers at each birth.

Decreasing lactation length is a simple way of increasing litter frequency if lactations are longer than six weeks. But not so for shorter lactations. Younger pigs are less able to cope with the ravages of life away from mother. The more severe the conditions, the less will the young pigs appreciate being weaned earlier. As lactation length is diminished then so must the housing, care and nutrition of the young be enhanced. When weaning at less than three weeks of age, the situation becomes so extreme as to demand a new technology in housing, husbandry, animal health and diet. Performances of European herds would suggest that the highest number of pigs produced yearly coincides with weaning at 21 to 28 days. This lactation length also requires least total feed per pig reared.

There is more scope for improving productivity by reducing the interval from weaning to conception. Even with six-week weaning and allowing only nine young pigs per litter, an annual production of 20 pigs per breeding female is possible if she is pregnant, as she ought to be, five days after weaning; while 17 pigs is more consistent with an average period between weaning and conception of 36 days. Infertility problems may be apportioned to the female, the male and to management. Each may need to be tackled at a level requiring considerable attention to detail.

The two most tractable aspects of productivity in the breeding herd are the interval between weaning and conception, and mortality. The resolution of problems in these areas rests with the individual husbandman into whose care the breeding herd is placed. And while technological information is essential to *enable* good husbandry, it does not necessarily bring it about.

Pig growth

3

To the pig, growth is the expression of an intrinsic need to reach mature size. To the pig producer, growth is the means to create a saleable product. The impulsion to grow comes from both nutrient supply and time (age); and pigs appear to have their own expectations as to the weight they ought to be by a certain age.

Mature weight most closely relates to the weight of the mature *lean* mass; fat content is notoriously variable in all adult animals.

Growth normally proceeds in a sigmoid or S-shaped manner (Figs 3.1 and 3.2). During early life the rate of weight gain accelerates, while between 30 and 120 kg growth is nearly linear. In later life, as maturity is approached, there is a decelerating phase to weight stasis.

The influence of time

The age of an animal when it becomes mature has considerable bearing upon the way it grows. Figure 3.3 shows some alternative growth curves for different pig types. Taking A as a reference against which to compare the others, B is an animal of lower mature weight but who reaches maturity at the same time; consequently B grows slower. C is of higher mature weight than A, but again reaches maturity at the same time; C grows faster. A positive association between growth rate and mature size is common amongst farm animals;

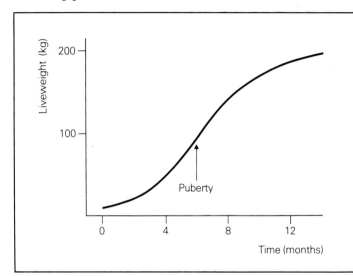

Figure 3.1   Growth curve for a pig. Weight against time

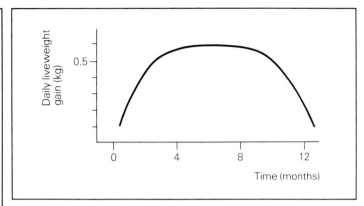

Figure 3.2   Growth curve for a pig. Gain against time

earlier and so grows faster without being larger. E is also of the same mature weight as A, but is later maturing and so grows slower.

compare for example the Aberdeen Angus and Charolais breeds of cattle, laying fowl and broilers, hill and downland sheep. It is likely that the same axiom applies to pig types; breeds and strains which have larger mature sizes tend to grow faster. Nonetheless, some heed should be given to possibilities D and E. D is of the same mature weight as A, but matures

Composition of growth

Live growth is comprised of the accumulation of lean tissue, fatty tissue and bone. Lean is mostly muscle and includes both carcass lean and lean in the offal and other body parts not normally eaten by man. Fat is mostly in the form of fat storage

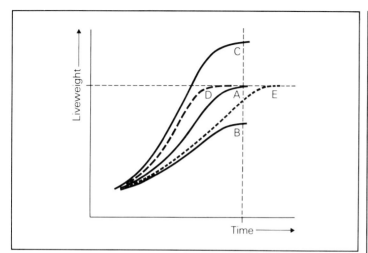

Figure 3.3 Growth curves for different animal types

depots under the skin (about two-thirds of total fat is subcutaneous) with some other fat accumulations occurring between the muscle bundles and around the kidneys and intestines. Bones have calcium phosphate as the major mineral together with fat, lean and connective tissues. About 10 per cent of the fat-free body is bone and the relationship between the quantities of bone and muscle in the body is fairly strict, bone being the support structure for muscle.

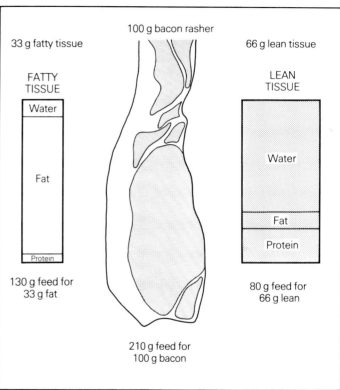

Figure 3.4 Composition of lean and fatty tissue

The chemical composition of lean varies, but is around 60–75 per cent water, 5–15 per cent fat and 20–25 per cent protein. In particular, the lean of fatter pigs contains more fat, and the lean of younger pigs contains more water. Fatty tissue has about 2 per cent of protein in it, but only about 8–12 per cent water. It is the difference in water content between lean (70%) and fat (10%) which makes lean so economical a tissue to produce (Fig. 3.4).

As pigs grow they get fatter (Fig. 3.5). The composition of pigs is primarily a result of the rate at which fatty tissue accumulates, and this depends on the amount of food the beast is consuming and the stage of maturity. Pigs get fat when there is nowhere else for food to go. The rate of fat growth increases with age and weight because as the animal becomes bigger it eats more, and there are excess nutrients over and above the requirements of lean growth. Fattening during the growing phase depends upon the amount of food supplied; the more food, the more fat will accumulate (Fig. 3.6). Later, as the pig approaches maturity, the rate of lean growth slows down, thereby again increasing the proportion of feed nutrients available for fat growth. The age at which fattening with the onset of maturity begins depends upon pig type; early maturing animals will fatten at younger ages.

Fat growth

Fat has different physiological significance at different stages

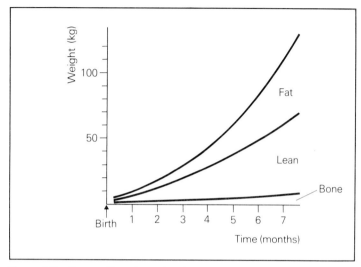

Figure 3.5   Composition of growing pigs

of life (Fig. 3.7). When born, the baby pig has only about 1 per cent of fat (Fig. 3.8), but he has a burgeoning impulsion to accumulate fat, his mother's milk, containing 8 per cent of milk fat, encourages this. With the first month of life the percentage of fat in the body has risen to around 15 per cent, and the pig will have attained the conventional, comfortable, rounded appearance (Fig. 3.9). This build-up of fat is no bad thing, for at weaning the combined effects of stress and a change of diet usually result in diminished food intake. The fat in the body is used as an energy reserve and is depleted by about half, down to 5 to 7 per cent of the body weight (Fig. 3.7); the severer the effect of weaning, the greater the fat loss. The change in the shape of the piglet bears witness to this (Fig. 3.10). When appetite returns, the pig readjusts to positive impulsion toward a level of fatness of around 10 per cent of the body weight. Once this level of *target* fatness has been reached, fatty tissue gain is no longer a priority, and the rate at which it accumulates flattens off (Figs 3.7 and 3.11). When food supply exceeds the requirements for the growth of lean, target fat and bone, fat storage depots are amassed and the fat level in the body rises *pro rata* with the level of feeding. At around six months of age the fat level may well be up to 25 per cent of body weight (Fig. 3.12), a condition not inappropriate for a female prepared to stand the rigours of motherhood (although grossly over-fat females may well experience difficulty in breeding). The fatness of the animal during its reproductive life varies with circumstances of housing, nutrition and management. The example in Fig. 3.7 shows a female losing about 12 kg of fatty tissue (equivalent to 7.5% of body weight) during lactation, and managing to claw back about 4 kg of fatty tissue during pregnancy.

The breeding female burns up her body reserve of depot fat, lactation by lactation, until a level of 5 per cent or less of fat in the body can be reached (Fig. 3.13). At this level of fatness the

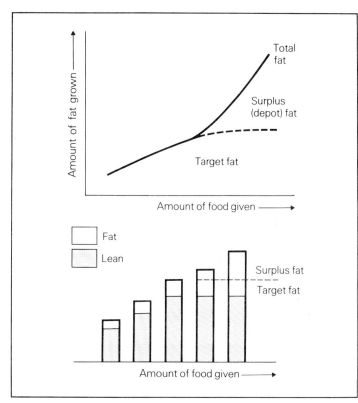

Figure 3.6 Influence of food intake on fat growth

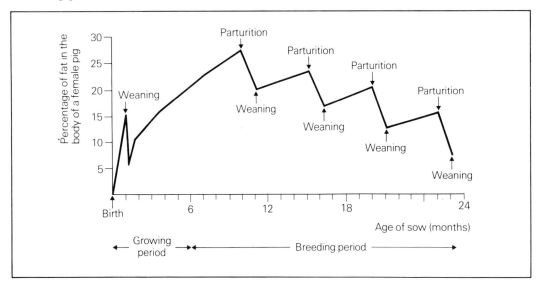

Figure 3.7 Change in fat composition of the body of a female pig. There are wide differences in fat composition of different pig strains. This picture represents conventional types. Modern lean strains will not achieve such high levels of fat at 6 months and may well never be above 15-20 per cent fat in the body. If the pattern of fat reduction shown here from 9 months is followed with modern lean strains, then problems of extreme thinness could well be exhibited prematurely

female is likely to encounter problems of reproductive failure. Decline of fat stores throughout breeding life is normal, but by no means compulsory, and in cases of overfeeding the body can have accumulated up to 50 per cent of fat.

Exploiting growth necessitates a distinction between *target* fat and depot fat. Target fat is a physiological necessity, while depot fat is a means of conserving a current food surplus in preparation for a future food shortage. It seems possible that pig types might differ in their levels of target fatness. Thus, fatties are impelled to reach a higher level of fatness than thinnies. Such differences are apparent between the sexes, and there may well be genetic differences between strains. In view of the demand for carcasses of low fat content, and the nutritional expense of fat growth as against lean growth (Fig. 3.4),

Figures 3.8-3.13 Changes in fatness of pigs from birth to old age

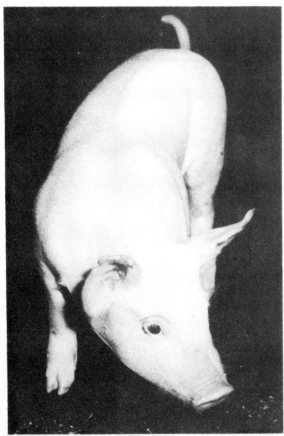

Figure 3.9

it would seem that the thinny is a pig of some merit; particularly as any extra fat needed can always be put on to the animal's back as depot stores.

On average, target fat is satisfied if the composition of the growth is about 20 per cent fat; that is, has a fat : lean ratio of 0.25 : 1.

In early life the nutritional environment should favour accumulation of fat and minimise fat depletion around weaning. Once target fatness is reached, it is in the interest of carcass quality to control the rate of growth of depot fat stores and this can be done by restricting the pig's ration allowance. Fat growth above target fatness is a direct consequence of the rate of feeding; more food produces more fat (Fig. 3.6). Apart from the reduction in the efficiency of food use which occurs when fat is produced rather than lean, carcasses will be downgraded if they are over-fat. Although it is probable that animals must attain the minimum target level of fatness for physiological reasons before their breeding life can begin, this level is likely to be exceeded by pigs in most commercially acceptable environments. The possibility of some pigs coming to puberty over-fat should not be taken as any justification for a *general* policy of slimming down all females between selection from the fattening house and mating at 100 to 120 kg. On the other hand, for females to enter the breeding phase in good body condition with 25 to 30 per cent of fat is an essential feature of a feeding regime which assumes a reduction in body stores throughout adult life.

Figure 3.10

Figure 3.11

Figure 3.12

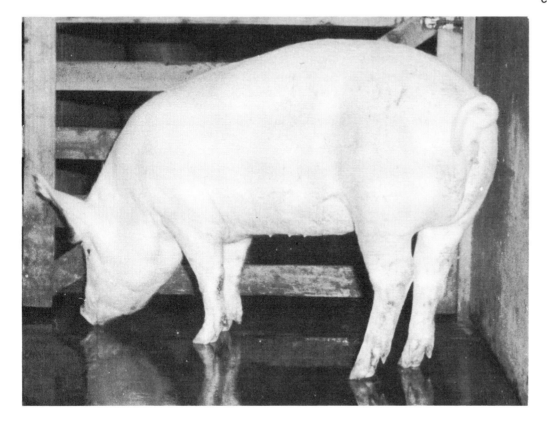

Figure 3.13

Lean growth

The pig is impelled towards maturity primarily by the growth of lean tissue. Lean does not, however, invariably have precedence over fat growth; in the young pig striving to achieve target fatness, lean growth is held back below its potential while nutrients are diverted to synthesis of fatty tissue. Lean growth is of major concern to pig producers because lean meat, not fat, is the product demanded by the consumer, and because lean growth requires only one third the food as compared to the same weight of fat growth.

While it is self-evident that lean growth makes lean meat, there is often confusion between lean growth and leanness, leanness being the expression of the ratio of lean : fat in the carcass, and therefore being a function of fat as much as of lean. Figure 3.14 indicates how rapid lean growth can be associated with both 'lean' (Case 1) and 'fat' (Case 2) carcasses, depending on the rate of fatty tissue gain. A lean carcass can also be produced from a pig which is not growing lean very rapidly (Case 3). Notwithstanding, it is indisputable that the dual objectives of fast growth and lean carcasses can only be achieved simultaneously if the pig grows lean at the maximum possible rate.

The maximum, or limit, rate which it is possible for a pig to grow lean will depend upon the sex and the genetic merit of the animal. Thus, females can grow lean faster than castrated males, and entire males can grow lean fastest of all. Improved strains of pigs selected for fast growth and lean carcasses are also likely to have a higher limit to their potential lean growth (Fig. 3.15). The increase in weight of the lean mass tends to follow a sigmoid growth curve similar to Fig. 3.1; as a result, daily lean growth rate appears to be as in Fig. 3.16. The flat top to the curve has been interpreted as indicating the maximum attainable rate. The decline in lean growth after 120 kg or so is undoubtedly a symptom of the approach of maturity.

The increasing phase of growth to the left of the picture in Fig. 3.16 is of particular concern because it relates to at least half of the productive life of pigs destined for meat production, and it seems that pigs are often unable to reach their maximum rate for lean growth until they are about 50 kg.

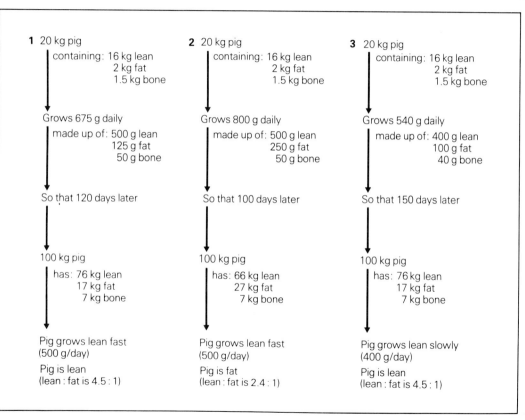

1 20 kg pig

    containing: 16 kg lean
                2 kg fat
                1.5 kg bone

Grows 675 g daily

    made up of: 500 g lean
              125 g fat
              50 g bone

So that 120 days later

100 kg pig

    has: 76 kg lean
        17 kg fat
        7 kg bone

Pig grows lean fast
(500 g/day)

Pig is lean
(lean : fat is 4.5 : 1)

2 20 kg pig

    containing: 16 kg lean
                2 kg fat
                1.5 kg bone

Grows 800 g daily

    made up of: 500 g lean
              250 g fat
              50 g bone

So that 100 days later

100 kg pig

    has: 66 kg lean
        27 kg fat
        7 kg bone

Pig grows lean fast
(500 g/day)

Pig is fat
(lean : fat is 2.4 : 1)

3 20 kg pig

    containing: 16 kg lean
                2 kg fat
                1.5 kg bone

Grows 540 g daily

    made up of: 400 g lean
              100 g fat
              40 g bone

So that 150 days later

100 kg pig

    has: 76 kg lean
        17 kg fat
        7 kg bone

Pig grows lean slowly
(400 g/day)

Pig is lean
(lean : fat is 4.5 : 1)

Figure 3.14 Lean growth and leanness. Three cases to show their independence. Quantities of lean, fat and bone refer to amounts in the total live animal body, not merely in saleable carcass joints (see Figure 1.5)

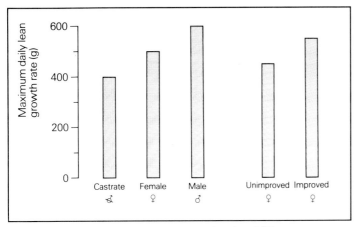

Figure 3.15   Potential daily lean growth rate for pigs of different sex and strain

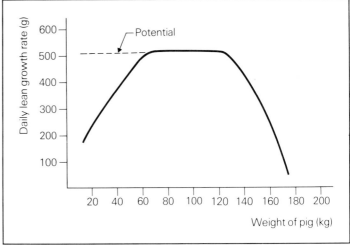

Figure 3.16   Daily lean growth rate of pigs in relation to live weight

Why this should be so is currently under hot dispute. It is possible that as the pig becomes older and bigger, the ceiling for maximum rate of lean growth actually rises. There is logic in this, as it is naive to envisage a 5 kg pig growing at the same rate as a 50 kg pig. Nonetheless, well cared for piglets in our Edinburgh Unit have grown 400 g of lean daily – which cannot be so far below their maximum – with little difficulty when only 15 to 20 kg liveweight. The other possibility is that appetite is an obstacle to maximising lean growth. This suggestion is apposite, as in early life there is the added dimension of a priority for target fat, with the diversion of nutrients away from lean growth. So a combination of a small stomach, the appetite-reducing traumas of weaning, and a prior claim on nutrients by fatty tissue, all conspire together to limit lean growth to a level below the potential.

The relationship between the composition of growth and food supplied to 60 kg pigs is shown in Fig. 3.17. The

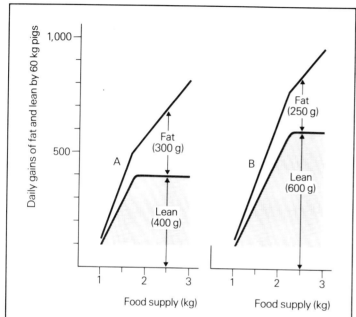

Figure 3.17   Daily gains of fat and lean by pigs in response to the rate of food supply. Pig A has a limit rate of lean growth of 400 g daily while pig B has a limit of 600 g. The boundary line gives the total gain; shown at 2.5 kg food supply to be 700 g for pig A and 850 g for pig B

consequences of this figure are fundamental to pig production strategy. In the example, pig A increases his lean growth rate in response to food intake up to a ration allowance of 1.75 kg. Any feed allowance of less than 1.75 kg is failing to cash-in on potential lean growth. Meanwhile, fat growth has kept pace with lean, commensurate with the need to maintain target fatness. The ratio of fat : lean of 0.25 : 1 satisfies the pig's requirement for target fat. Because of the pig ever striving to reach maturity, there is an internal drive for the pig to accumulate lean as fast as possible. This drive to maximise lean growth means that the pig cannot get fatter than target fatness dictates, just so long as food is restricted at a level below that which will satisfy maximum lean growth together with target fat. It would appear that average pigs restricted to a ration which does not allow any accumulation of fat above the target level will, at 90 kilo, have P_2 backfat measurements in the region of 12 to 15 mm; rather less for animals with lower target fatnesses.

When maximum lean growth is reached (at 1.75 kg food supply in Fig. 3.17 A) increments in food supply cannot enhance lean, but go towards fat. Target fat is exceeded and fatness relates to the excess of feed over the requirements for lean. The total growth rate response to food is broken downwards at this point on account of the different efficiencies of food use for lean and fat growth.

Although no difference is evident between pig A and pig B until A reaches his maximum lean growth rate, pig B is an

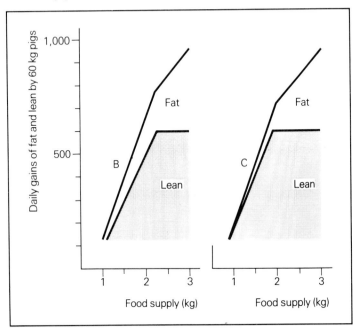

Figure 3.18   Daily gains of fat and lean by pigs in response to the rate of food supply. Both pigs have limit rates of 600 g lean growth daily, but while pig B demands fat growth at a minimum fat:lean ratio of 0.25:1, pig C demands a minimum fat:lean ratio of 0.125:1. The boundary line gives total gain (at 2.0 kg food supply; 650 g for pig B and 700 g for pig C)

animal of greater potential lean growth and needs more food (up to 2.25 kg) to satisfy that potential. B can take more food and grow faster without becoming fat and inefficient, as may be confirmed by a comparison of the performance of the two pigs when both are given 2.5 kg of food.

Pigs differ in the level of physiological, or target, fat that they demand. This will further influence the response to food supply (Fig. 3.18). Pig C is a thinny, evidenced by a fat : lean ratio of 0.125 : 1 satisfying the minimum requirement for fat, rather than 0.25 : 1 as for pig B. The lower rate of target fat accumulation results in pig C growing lean faster (reaching maximum lean growth at 1.9 kg food supply) and also being leaner and more efficient at levels of food supply below 2.25 kg (the point at which pig B maximises his lean growth). At levels of food at which both pigs are maximising lean, their growth patterns are identical. There is some reason to believe that entire male animals and improved strains of pigs are thinnies, while castrated males and unimproved pigs are fatties.

Diet selection

4

Diets are made from mixtures of feed ingredients. The pig can accept a wide range of feed types, but digests roughage poorly. Feeds high in fibre and of low nutrient concentration may limit food intake. Most diets are based on mixtures of cereals (barley, wheat, maize, oats) with some protein-rich soyabean meal or animal and fish by-products added. Other ingredients (tapioca, potato, milk products, fats, roots, sorghum, millers' offals and other vegetable protein sources) also find their way into pig diets, as do medicaments and chemical growth promoters.

Feeding pigs is the major activity on the unit, and feed the major expenditure. Feed can only be used efficiently if the nutrients in it are balanced to satisfy the animal's specific needs. Pigs have different nutrient requirements according to their age and weight, and depending upon the rate and type of productivity required of them. Diets therefore need to be chosen circumspectly.

The strict meaning of diet is different from ration. Diet implies the type of food; ration the amount. It is absolutely impossible to ration pigs properly without intimate knowledge of the diet concerned. The pig producer may purchase diets from feed manufacturers, or he may compound them for himself. In either case sound feeding programmes require at least three pieces of information about diets.

1. The energy concentration.
2. The protein concentration (or energy : protein ratio).
3. The potential utility of the protein (or biological value).

Vitamins and minerals, so fashionable in the human diet, although equally essential to the pig, are not particularly interesting in production terms. Low cost and indispensibility should ensure their inclusion into diets *de rigueur* at or above requirement levels. It is the dietary concentrations of energy and protein which control the productivity of breeding stock, determine the rate of gain of growing pigs and have a major influence upon carcass quality.

Dietary protein mostly goes towards the synthesis of lean growth, into which protein is incorporated at the rate of about 22 per cent. A little dietary protein is also used for purposes of body maintenance. Any excess protein eaten by the animal, but not required for these purposes, is broken down by the body; the ammonia goes to waste as urea in the urine, and the remainder adds to the energy supply.

Dietary energy is incorporated into both fat and lean growth and, most importantly, provides fuel to drive the metabolic functions of the body. The body uses energy to make lean, to make fat and to maintain itself. Maintenance represents the day-by-day costs of keeping the pig alive, healthy, active and growing. Between a third and a half of all energy given to pigs is used for maintenance, and in this respect it is non-productive; the energy used does not appear in any saleable product. Maintenance needs are strictly related to the size of the animal; the larger the animal the more energy it uses up.

The concentration of dietary protein is usually given as grams of crude protein or digestible crude protein per kilo of feed (g CP or g DCP/kg), the biological value of the protein as a fraction of unity, and the concentration of energy as megajoules of digestible energy per kilo of feed (MJ DE/kg).

Dietary concentration of energy (energy density)

Energy concentration may range from around 11 MJ/kg or less for diets containing grass meal and other roughages, to 15 MJ/kg for diets with maize and added fat. In general, energy concentration is positively related to dietary fat (L) content and negatively related to the fibre (CF) content (DE can be approximated by use of the equation: DE $(MJ/kg) = 14.2 - 0.45 (CF\%) + 0.25 (L\%)$). Most conventional diets fall within the range 12.0 to 14.0 MJ DE/kg.

Naturally, the nutrient density of a given weight of diet depends on the water content. In most circumstances diets are prepared and rationed in dry form, and water added being incorporated at feeding time. Some excellent diet ingredients may, however, be available in the wet form, for example, root crops (10–25% dry matter), fish silage (15–25% dry matter) and liquid milk by-products (6–13% dry matter).

Cereal/soyabean mixes are usually 85 to 90 per cent dry matter (depending on the moisture content of the cereals), and the convention is to assume about 87 per cent of dry matter

(13% water) in a pig diet unless otherwise stated. Comparison of diets of different water content requires correction to equal water content (Table 4.1).

Table 4.1 Comparison of diets of differing water content

	'Wet' diet	'Dry' diet
Dry matter of diet (%)	32.8	87.0
DE content of diet (MJ DE/kg)	4.9	12.4
DCP content of diet (g DCP/kg)	50	123
DE content of diet corrected to 87% dry matter	13.0[a]	12.4
DCP content of diet corrected to 87% dry matter	133[b]	123

[a] $(4.9 \div 0.328) \times 0.87 = 13.0$
[b] $(50 \div 0.328) \times 0.87 = 133$

It is self-evident that to provide a pig with a given quantity of energy, more food needs to be offered if the diet is of low energy concentration, and less if the diet is of high energy concentration. To provide 30 MJ DE, 2.6 kg is needed of a diet with 11.5 MJ DE/kg while only 2.2 kg is needed of a diet with 13.5 MJ DE/kg. For rapid growth or high production where large nutrient intakes are required, more concentrated diets are indicated; particularly if the pig's capacity for food is limited, as is often the case with young growing pigs. In contrast, animals of lower productivity with big appetites, such as growing pigs above 60 kg and pregnant females, can utilize diets of lower nutrient density (Table 4.2). Any pig rationed to a feed allowance less than appetite could, if

Table 4.2 Lower limits for energy density range appropriate to various types of pig[a]

	MJ DE/kg
Baby pig (5–20 kg)	14
Grower (20–60 kg)	13
Grower (60–100 kg)	12
Pregnant breeders	< 11
Lactating breeders	12.5

[a] The upper limit for the energy density range is, of course, not dependent on the type of pig, but on the feasibility of the diet. Thus, all pigs could be given diets with 14 MJ DE if this was the most economic concentration (perhaps if maize was inexpensive compared to barley). Pigs offered higher density diets should have lower feed allowances *pro rata*.

required, be given a diet of lower nutrient density; but it is not always economic to do so.

Energy : protein ratio (MJ DE : g DCP)

Dietary protein may be measured as crude protein (CP) or digestible crude protein (DCP). Should the less useful CP term be the only one available, an approximation of DCP can be made on the assumption that DCP is about 85 per cent of CP. Such an approximation, of course, fails to allow for an actual range of 40 per cent to 90 per cent in digestibility between different dietary protein sources; it is for this very reason that DCP is so much the more useful value.

The concentration of protein required in the diet depends on protein needs for lean growth or milk production in comparison to energy needs for body fuelling. Large animals have high energy demands for body maintenance and therefore need high energy diets with narrow energy : protein ratios. Animals whose composition is to be low in lean and high in fat growth have a low protein requirement and therefore similarly do not need diets of high protein concentration. Smaller animals with lower maintenance needs, and animals with high lean and low fat growth rates need more protein, and suitable diets are likely to have wider energy : protein ratios (Table 4.3).

Energy : protein ratio divides diets into three broad classes. Diets with DE : DCP ratios narrower than 1 : 10 (DE : CP ratios narrower than 1 : 12) are appropriate to pregnant adult females and growers of above 80 kg liveweight. Diets with DE : DCP ratios wider than 1 : 10 (DE : CP ratios wider than 1 : 12) are appropriate to growing pigs of less than 40 kg. Diets with DE : DCP ratios of around 1 : 10 (DE : CP, 1 : 12) are appropriate to lactating females and growing pigs between 40 and 80 kg.

Diets are often described by their protein density (g DCP or g CP/kg diet), but this value is of little use unless the DE density (MJ/kg diet) is also known; that is to say unless protein content can be expressed as the DE : DCP or DE : CP ratio. Table 4.3 indicates that to provide for the pig's protein requirements, a diet of higher energy density may need also to be of higher protein density. The range of possible energy densities for pig diets is wide, the upper limit being about 15 MJ DE/kg, and controlled by the availability of high-energy ingredients; while the lower limit depends on pig type and may go lower than 11 MJ DE/kg, depending on the animal's appetite (Table 4.2). 140 g DCP (165 g CP)/kg represents a high level of protein in a diet with 12 MJ DE/kg, but a low level of protein in a diet with 15 MJ DE/kg.

The number of different diets found on a single pig unit depends upon the number of different pig types requiring either differing energy densities, or differing energy : protein ratios. On some units, simplicity of management outweighs nutritional efficacy, and a single diet of about 13 MJ DE and with a DE : DCP ratio of 1 : 10 (DE : CP, 1 : 12) is used for all pigs. Such an arrangement is straightforward, but is far from optimum in terms of efficiency of food use, and may cost money through lost production and wasted food. The use of as many diets as there are pig types will maximize the physical efficiency of food use, but the feasibility of such a programme will depend much on the size of the unit. An often held convention is to have a special diet for baby pigs, a diet for growers between 20 and 100 kg (or two diets: one for pigs below 60 kg and another for pigs above 60 kg) and a diet for sows (or two diets: one for pregnancy and another for lactation).

Table 4.3 Nutrient concentration of diets for different types of pig

	DE density[a] (MJ/kg)	CP density (g/kg)	DCP density (g/kg)	DE:CP[b] ratio	DE:DCP[b] ratio
Young grower (20 kg)[c]	14	200	170	1:14	1:12
Grower (40 kg)[c]	**13.5**	175	150	1:13	1:11
Grower (60 kg)	13	153	130	1:12	1:10
Grower (80 kg)	13	140	120	1:11	1:9
Grower (100 kg)	13	135	115	1:11	1:9
Pregnant breeder	12.5	140	120	1:11	1:9
Lactating breeder	13	153	130	1:12	1:10
Improved entire male grower (40 kg)	14	182	155	1:13	1:11
Unimproved castrated male grower (40 kg)	13	140	120	1:11	1:9

[a]An average expectation, see Table 4.2.
[b]Assuming a biological value of 0.65. Higher DCP (CP) levels are required if protein is of lower biological value.
[c]With young pigs palatability is as vital a part of diet quality.

Biological value

The value of feed proteins for pigs relates to their amino acid make-up as compared to the balance needed by the pig. A perfect match between amino acids in feed protein and amino acids needed is expressed as a biological value of unity. The proteins of some feed ingredients – for example, soyabean and fish meal – impart high biological value to a diet; while others – for example, the protein of cereals – are deficient in the amino acid lysine and thereby impart low value to a diet.

Biological value, like energy density, is not an objective to be sought after, but rather something that one needs information about, provided only that the lower limits given in Table 4.4 are observed. Although dietary protein of high biological value will be more efficiently utilized by the pig, it may not be cost effective compared to using *more* of a *cheaper* lower quality protein. Using a protein of lower biological value requires the diet to have a wider energy : protein ratio, and thus a higher DCP level (see also footnote *b* to Table 4.3).

Table 4.4 Lower limits for biological value of dietary protein appropriate to various types of pig[a]

	Biological value
Baby pig (5–20 kg)	0.70
Grower (20–60 kg)	0.65
Grower (60–100 kg)	0.60
Pregnant breeders	0.60
Lactating breeders	0.65

[a] The upper limit for biological value of dietary protein is dependent on the feasibility of the diet.

Where information on the biological value of diet protein is not available, then a statement of the dietary level of the amino acid lysine[1] can be of help; this should be readily obtained from the feed compounder. Because lysine is usually the amino acid which controls biological value, then with the knowledge that the pig's requirement is for 0.07 g lysine per gram protein, biological value can be calculated quite simply from the dietary lysine level:

$$BV = [\text{Lysine in diet (g/kg)}/\text{CP in diet (g/kg)}]/0.07$$
$$\text{or } BV = \{[\text{Lysine in diet (g/kg)} \times 0.85]/\text{DCP in diet (g/kg)}\}/0.07$$

[1] More strictly, *available lysine* is a better value for these purposes than total lysine. In most diets, but not all, lysine is about 90 per cent available.

Table 4.5 Expected characteristics of pig diets

	Energy density (MJ DE/kg)	DE:DCP ratio	DE:CP ratio	Biological value of protein
Baby pig (5–20 kg)	14–15	1:12–15	14–18	0.7–0.8
Grower (20–60 kg)	13–14	1: 9–13	11–15	0.65–0.75
Grower (60–100 kg)	12–14	1: 8–10	9–12	0.6–0.75
Pregnant breeder	11–13	1: 8–10	9–12	0.6–0.7
Lactating breeder	12.5–14	1: 9–11	11–13	0.65–0.75

Summary

Table 4.5 summarises the expected ranges for energy density, energy : protein ratio and biological value in pig diets. This information ought to be obtainable directly from the diet compounder, but might have to be derived by calculation. Characterization of diets in this way not only allows the producer to present the most appropriate diet to the various type of pigs, but also allows prediction of the likely outcome of a change in diet quality.

Response to a change in dietary density of DE

An increase in dietary DE concentration is appropriate where pig productivity is below potential despite the animal eating to maximum gut capacity, as is often the case for young growers. If the more concentrated diet ingredients are more expensive, then the extra productivity in terms of growth or milk must be weighed against the extra cost. For pigs whose appetites are greater than the productivity required of them, a decrease in DE concentration can lead to cost-saving; provided that the extra amount needed of the cheaper diet does not cancel out the benefit.

If the DE : DCP ratio remains unchanged then an increase in DE density will be exactly matched by a *pro rata* increase in DCP concentration. The effect upon the pig will therefore be just as if it had received more food. To maintain constant performance while changing to a diet of lower or higher energy density, the ration allowance must be increased or decreased respectively (Fig. 4.1). Conversely, energy supply will be enhanced if energy density is increased without any commensurate change in ration (Fig. 4.2). The pig responds to intake of energy, not weight of food.

Response to food supply made by 60 kg growing pigs has been described previously in Figs. 3.17 and 3.18. Increasing energy supply through an increase in diet density has the same effect. First, the rate of lean growth is increased until maximum daily lean growth is obtained. During this phase the pig is unlikely to get fat. After lean growth is maximized, further increases in energy supply go to fat growth. Pigs with higher lean growth potentials can use more dietary energy before running to fat. Figure 4.3 shows how the response to energy would relate to diets of differing energy density; 2.25 kg of diet with 11.5 MJ DE/kg would fail to maximize lean growth, while the same ration of a diet with 14.5 MJ DE/kg

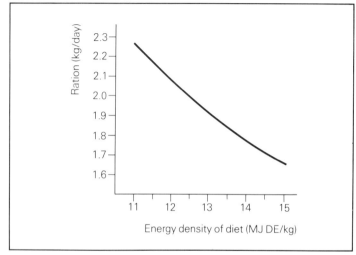

Figure 4.1   Rations required for diets of different energy concentrations to give the same response of 625 g live weight gain on a 60 kg pig (500 g lean and 125 g fat)

would actually exceed the requirements for lean and target fat.

Responses of adult females to a change in dietary energy concentration are as predictable as those of the growing pig. If the change in DE is accompanied by an equivalent and contrary alteration to the ration supply, such that nutrient intake remains unaltered, then there will be no change in

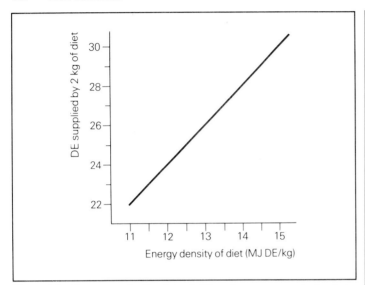

Figure 4.2 Energy supplied by diets of different energy concentration fed at the same ration allowance of 2 kg daily

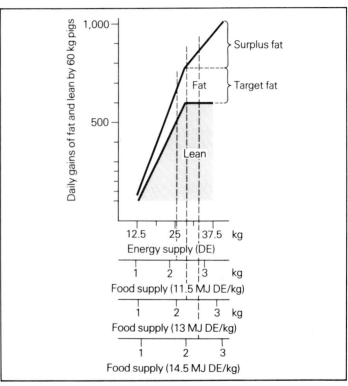

Figure 4.3 Daily gains of fat and lean for a pig given diets of differing energy density. The broken lines compare responses to a ration of 2.25 kg

productivity. If the ration level (or appetite intake) remains stable while DE alters, then the animal will respond to a decrease in energy supply by becoming thinner and to an increase by becoming fatter. In the case of females already thin, pregnant females given more energy will produce heavier young, lactating females more milk and weaned females will have improved conception rates. Overfat sows, apart from being a needless waste of feed, are also less productive, and a slimming programme would certainly reduce feed costs and might even enhance output.

Response to a change in dietary DE : DCP (DE : CP) ratio

Provided that lean gain is not held back by insufficient dietary energy, then increasing the supply of dietary protein will increase lean growth until the maximum potential is attained. Additional increments of dietary protein cannot raise lean growth above the maximum, while animals with higher potential can make use of more protein.

It cannot be assumed that pigs not already maximizing lean growth would do so if given more protein; for this would be to assume that lean growth was limited by a dietary protein deficit, and this is by no means always the case. As shown in Fig. 4.3, lean growth may be just as readily limited by an energy deficit.

Figure 4.4 illustrates the type of growth responses of a 60 kg pig given either of three rations of diets all with 12.5 MJ DE/kg, but with energy : digestible protein ratios between 1 : 8 and 1 : 12 (DE : CP ratios between 1 : 9 and 1 : 14).

At a ration level of 1.5 kg daily (A), a diet with a DE : DCP ratio of 1 : 8 does not supply sufficient protein for lean growth. Increasing the protein level to a DE : DCP ratio of 1 : 10 corrects the situation and lean growth rate improves. Dietary energy and protein are in balance at this point and there is just sufficient energy to fuel both target fat and the lean growth rate allowed by the level of dietary protein. A further increment of protein to a DE : DCP ratio of 1 : 12 does nothing to alter the position, and cannot further enhance lean growth; lean growth is limited by an energy deficit. The excess protein is, in fact, an embarrassment because it must be broken down, which itself uses up precious energy. The DE : DCP ratio of 1 : 10 maximizes lean growth and minimizes fat growth only within the confines of the ration level, which in case A is meagre. At best, lean growth is well below the potential.

Increasing the ration to 2.0 kg (B) brings about a general improvement in performance level. But at DE : DCP ratios of less than 1 : 10 there is again insufficient protein to maximize lean growth, and there is excess energy, which induces fatness. Maximum lean is finally achieved when 2 kg of diet with ratio of 1 : 10 is given; this diet also minimizes fat deposition. At DE : DCP ratios above 1 : 10, lean growth is limited by the pigs potential (in case A it was limited by energy shortage), but the animal will not get over-fat because there is no excess energy. In contrast, with a ration level of 2.5 kg (C) there is excess energy available, and this causes the accumulation of surplus fat storage at all protein levels.

Figure 4.4 demonstrates that if the DE : DCP ratio is too

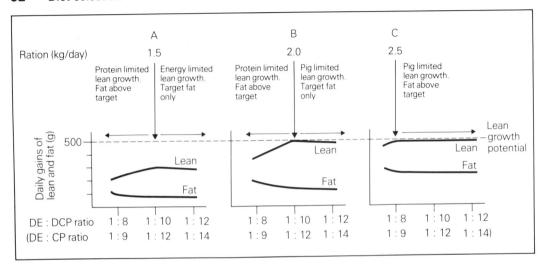

Ration (kg/day)

A — 1.5
- Protein limited lean growth. Fat above target
- Energy limited lean growth. Target fat only

B — 2.0
- Protein limited lean growth. Fat above target
- Pig limited lean growth. Target fat only

C — 2.5
- Pig limited lean growth. Fat above target

Daily gains of lean and fat (g)

500 — Lean growth potential

Lean
Fat

| DE : DCP ratio | 1 : 8 | 1 : 10 | 1 : 12 | 1 : 8 | 1 : 10 | 1 : 12 | 1 : 8 | 1 : 10 | 1 : 12 |
| (DE : CP ratio | 1 : 9 | 1 : 12 | 1 : 14 | 1 : 9 | 1 : 12 | 1 : 14 | 1 : 9 | 1 : 12 | 1 : 14) |

Figure 4.4 Influence of DE:DCP ratio on daily gain. In the examples a 60 kg pig with a lean growth potential of 500 g daily is offered diets of varying DE:DCP ratios at 3 levels of daily ration: A, 1.5 kg; B, 2.0 kg; C, 2.5 kg

narrow, lean growth is retarded. Widening the ratio has a direct and positive effect upon the rate of lean growth, and only when the correct DE : DCP ratio is attained can lean growth be maximized. Fatness is readily induced by inadequate protein supply, and this too can be corrected by widening the energy : protein ratio. Nonetheless, it is also unequivocally demonstrated that pigs can grow surplus fat at any DE : DCP ratio if the ration level is greater than the pig requires. Excess protein cannot enhance lean growth above

either the limit set by available energy or by the inherent potential of the pig.

The lactating pig can produce 400 to 500 g of protein in her milk daily. Milk production in the adult is a similar body process to growth in the young animal, requiring copious amounts of both protein and energy for incorporation into milk protein, milk sugars and milk fat. Additional energy is also required to fuel the synthetic processes, body maintenance and any need there may be for the accumulation of fat depot stores. A reduction in the DE : DCP ratio below the 1 : 10 (DE : CP, 1 : 12) mark is likely to cause a reduction in milk yield and an increase in the rate of lean tissue loss from the mother's body. Widening the ratio will correct any protein deficiency, but a (DE : DCP) ratio of above 1 : 11 (DE : CP, 1 : 13) is not likely to have any further positive effect upon milk yield. Just as in the growing pig, response to DE : DCP ratio is influenced by the general level of energy supply. Inadequate food provision will fail to maximize milk yield regardless of protein level.

The pregnant animal is relatively frugal in its protein needs. It is not lactating, neither is it growing lean particularly actively. Increasing diet protein concentration may slightly increase liveweight gain during pregnancy, but within the normally feasible range (DE : DCP, 1 : 8–14; DE : CP, 1 : 9–17) will have no effect on number of pigs nor on their birth weights. The needs of the foetuses are very modest until the last 20 to 30 days of pregnancy. It has been proposed that a DE : DCP of 1 : 8 (DE : CP, 1 : 9) is adequate for the pregnant female. But there is some evidence that low protein levels may reduce fertility (both conception rate and numbers born), particularly if the protein is of low biological value. With this possibility in mind, some producers adhere to a policy of feeding the diet prepared for lactating breeders to weaned females until they are confirmed as being pregnant again. As mentioned in Chapter 2, it is regrettable that so little is known of the factors influencing reproductive performance, but as it is so, the influence of diet on the reproductive processes remains highly speculative. There is again nothing that the animal can do with a superfluity of protein, and on the basis of pragmatism, a DE : DCP ratio of around 1 : 9 (DE : CP, 1 : 11) is probably what is required for pregnant breeders.

Response to a change in biological value

Biological value governs the effective utilizability of the dietary DCP. A change in biological value therefore changes the effective DCP content of the diet. Thus, a reduction in biological value can be countered by an increase in DCP. It follows that response to change in biological value is similar to the response to change in DCP, an alteration to biological value causing a proportional alteration in effective DCP supply.

The protein level chosen for any diet must allow for biological value of the proteins used in the mix, or errors will arise from a DCP or CP supply being assumed adequate when, in fact, it is not.

Ration selection

5

The feed allowance, or ration, is the control valve for both rate of output and product quality in growing and breeding pigs. Rationing is therefore a prime determinant of pig performance and profitability. At one extreme, feed restriction can cause growing pigs to be so lean that their carcasses are hardly edible, while breeding females can lose body condition and become irredeemably emaciated. At the other extreme, appetite-fed pigs can bowl along at the rate of a kilo-a-day toward gluttonous obesity. The art in rationing pigs is to pitch at exactly the right spot: to grow as fast as possible without reducing carcass value through fatness, and to build up body reserves of breeding females without being profligate (Fig. 5.1). The right spot along the scale of rationing between severe restriction and appetite feeding is elusive, shifting with market forces and differing for each producer's individual circumstances.

Optimum rationing requires first the selection of the appropriate *type* of scale, and then choice of the *level* at which the scale is to be set.

Types of ration scales

Although there are many types of ration scales, they are all directed toward the practical end of providing a programme to control the frequency and size of the feed increments required by pigs as they grow, or as they proceed through their

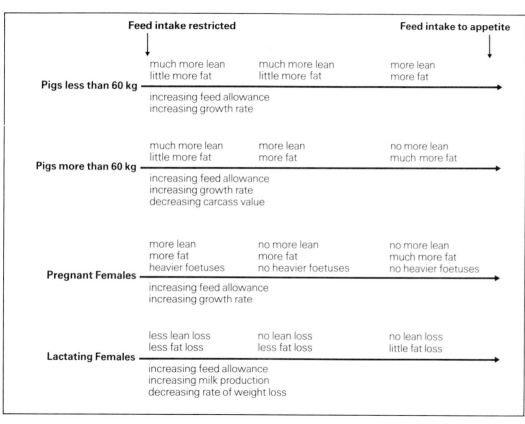

Figure 5.1 Influence of increasing feed allowance upon pig productivity

breeding life. Because the amount of nutrients required for maintenance depends on body size, then even to grow at the same rate, bigger pigs need more feed than smaller pigs. While independent of size, greater productivity of both growers and breeders necessitates the provision of a higher feed allowance.

The upper limit to the amount of feed a pig can consume is governed by appetite. Appetite is a complex characteristic, varying at the least with pig type, farm, season, temperature, housing, food texture, palatability of diet ingredients and method of feeding. But whatever the appetite level, it will have been constrained either by the capacity of the pig's gut (Fig. 5.2) or by self-regulation of total energy input (Fig. 5.3). Gut capacity is an important limit on productivity of pigs given diets of low nutrient density, and of pigs with high potential growth rates. Energy limits often apply when pigs are given higher density diets; stopping them eating more, despite unused physical capacity.

The lower limit to feed intake is the amount required for maintenance (Fig. 5.4), for if less than this is given lactating

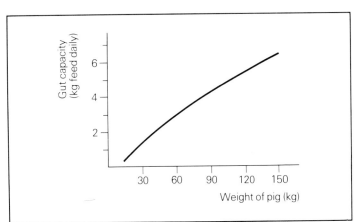

Figure 5.2 Appetite feed intake limits (gut capacity), under average commercial conditions

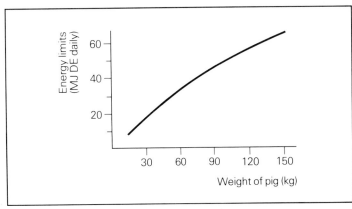

Figure 5.3 Appetite feed intake limits (energy), under average commercial conditions

females will rapidly shrink, while growers and pregnant pigs will gradually fade away.

As young growers have a high potential for lean growth in comparison to appetite, the problem is usually one of *under-eating*. Not only is feeding to appetite the appropriate ration scale, but appetite is often unnecessarily depressed by ill-health, trauma, unsuitable housing and inadequate self-feeding methods resulting in competition over stale feed. The only circumstance where a young pig's intake may need to be curtailed rather than enhanced is on a unit where diarrhoea is a problem; strict rationing can reduce the severity of the symptoms.

From about 30 to 60 kg onwards (depending on the quality and sex of the pig; see also Fig. 3.15) rationing scales of various degrees of restriction may be imposed. There are broadly two types of scale: those which allow pigs to eat to appetite but have a maximum limit point after which a flat rate is fed (Fig. 5.5) and those which provide an allowance related to the increasing weight of the pig (Fig. 5.6). Both types provide similar allowances in early growth, but scales with maximum limit points, while easiest to implement, provide in reality a decreasing allowance to the heavier pigs,

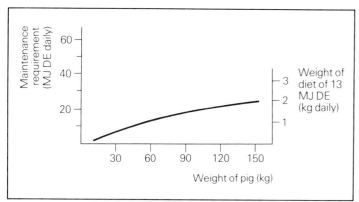

Figure 5.4 Maintenance requirements

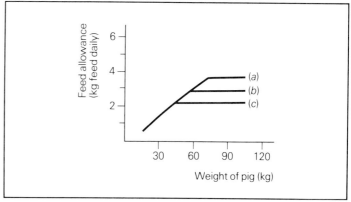

Figure 5.5 Ration scales which allow appetite intake up to a maximum limit. (*a*), (*b*) and (*c*) represent increasing severity of feed restriction

the flat-rate making no allowance for the increasing needs of maintenance as animals grow. Weight-based scales may be operated on the basis of actual weight or, as is more practical, estimated weight. A ready guide to weight is the pig's age; specifically, the time he has spent on a known feed intake. This is only reasonable as liveweight gain is the consequence of food supply and not vice versa. Time-based scales (Fig. 5.7) are probably the most satisfactory type. They are simple to operate, as well as capable of relating closely to the needs of the growing pig. A further advantage of the time-based scale is that, unlike a weight-based scale, no cognisance is taken of transient interruptions to growth. If the growth rate of a group of pigs falls behind a little, the feed allowance is proportionately more generous, which results in the growth path being smoother and more predictable. Time-based scales can be incremented weekly, fortnightly or monthly. The more

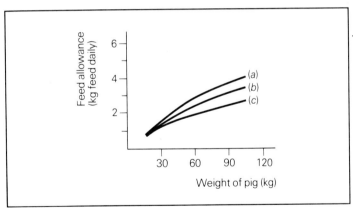

Figure 5.6   Ration scales related to the weight of the pig. (*a*), (*b*) and (*c*) represent increasing severity of feed restriction

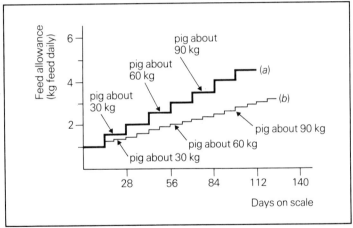

Figure 5.7   Ration scales related to time. (*b*) represents a more severe restriction than (*a*). Scale (*a*) is incremented every 14 days, and scale (*b*) every 7 days. The scales shown are started when pigs are about 20 kg. Conventionally, scales such as these would be limited to maxima of 2.25–3.5 kg

frequent the ration adjustment, the more accurately the pig's requirements are matched.

A combination of scales is commonly used. Thus, pigs may eat to appetite until they reach a certain weight, or until they are moved to different accommodation. Then a time-based scale is imposed for a predetermined duration until a maximum limit point is reached, after which no further increments are allowed.

Pregnant females are conventionally rationed to a flat rate, the amount depending on body condition and environmental circumstance. The greatest demands of pregnancy come in the last quarter, and it has been suggested that an increase in ration allowance at this time might enhance birthweight and improve piglet survival rate. There is little solid evidence to substantiate this. Between weaning and conception, and no bad thing until the third week of pregnancy, a higher feed level is, however, appropriate (Fig. 5.8) to encourage the greatest possible litter size.

After parturition, lactating animals need three to four days to be built up to their full feed allowance. The final ration is usually given as a flat rate during the course of the lactation

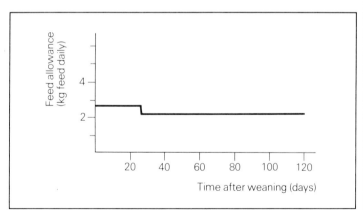

Figure 5.8   Ration scale for a pregnant female

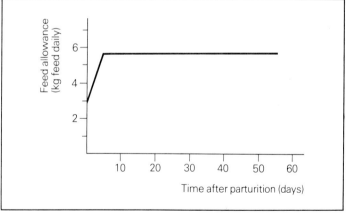

Figure 5.9   Ration scale for a lactating female

(Fig. 5.9). There is no benefit from attempting to feed to any lactation curve. The amount may be fixed for all females, or may relate to the required average level of milk production for individual animals.

Responses of growing pigs to a change in ration level

The appropriate level at which to pitch a ration is not a matter open to general recommendation. It depends entirely upon the particular responses that individual producers require of their pigs, the objective being to optimize financially within the peculiar conditions of the prevailing market forces as they apply at specific times and at specific geographical locations.

Increasing the feed allowance causes pigs to grow faster and encourages fat deposition (Fig. 5.10). The instantaneous position at 20, 60 and 100 kg is depicted in Fig. 5.11. Whilst more food always results in more fat, with young pigs more lean is also accumulated, resulting in faster growth and improved feed conversion efficiency. By 60 kg the normal food intake range spans the point at which extra food goes to surplus fat rather than lean; the rate of growth response to increasing food slows, and feed conversion stabilizes. Comparison of the three weights of pig in Fig. 5.11 demonstrates two fundamental aspects of the growth response to increasing food supply. First, as pigs grow they need more

food to achieve the same output; 600 g of lean and 150 g of fat are achieved with an intake of 2.25 kg at 60 kg, but need 2.5 kg at 100 kg. Second, increasing pig size brings with it increasing appetite but not increasing ability to grow lean. The chance of overfeeding pigs and accumulating too much surplus fat therefore increases with pig age.

Above all, fast growth, particularly fast lean growth, results in considerable energy savings by reducing the food drain to

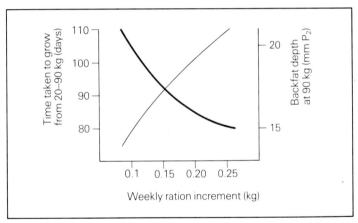

Figure 5.10   Influence of feed allowance on backfat depth (mm P_2) and daily live-weight gain (days from 20–90 kg). Each scale started at 1 kg when pigs were 20 kg

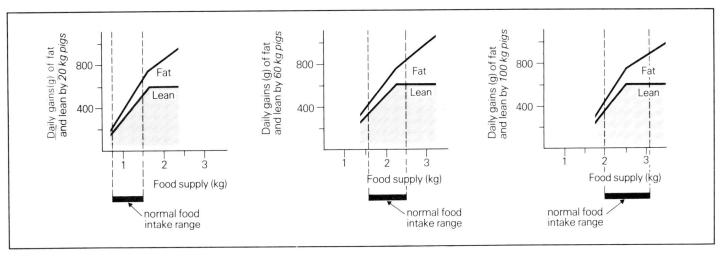

Figure 5.11 Daily gains of fat and lean in response to rate of food supply

maintenance. The faster the pig grows the sooner slaughter weight is reached. As maintenance is a daily expenditure, fewer days alive means commensurate food savings. Each extra day spent by a 70 to 80 kg pig in getting up to weight costs in the region of an extra kilo of feed.

The influence of the incremented ration scale becomes increasingly important for pigs slaughtered at heavier weights (Fig. 5.12). Between the lowest and highest feed allowance

shown there is about 1.5 mm difference in backfat at 60 kg, while at 120 kg the difference is nearer 10 mm. But perhaps the major forces qualifying response to ration change are the characteristics of the pig – as typified by genetic strain or sex (Fig. 5.13), disease, and environment – particularly house temperature (Fig. 5.14).

Often, increments suitable to keep up with the expanding needs of the young pig are too great for older pigs, for whom

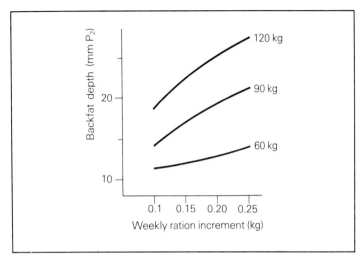

Figure 5.12 Influence of feed allowance on backfat depth at 60, 90 or 120 kg. Each scale started at 1 kg when pigs were 20 kg

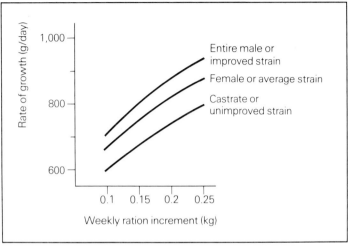

Figure 5.13 Influence of feed allowance upon average rate of growth from 20–90 kg for pigs of different inherent growth potential

the increment is required only to keep pace with maintenance. Ideal ration scales should perhaps increase by steps of decreasing size. This is tacitly accepted in the type of practical scale which allows pigs first to eat to appetite up to a given liveweight, next restricts the animals to a time-based scale starting at an assumed intake and last settles to a flat rate at some maximum point of the time scale (for example, to appetite to 40 kg, then 1.8 kg raised by increments of 0.15 kg weekly over six weeks to a maximum of 2.7 kg which is given as a flat rate until slaughter at 90 kg).

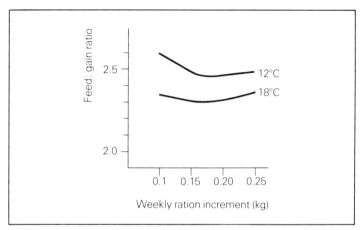

Figure 5.14 Influence of feed allowance upon efficiency of food use (feed: gain ratio) from 20–60 kg for pigs at different house temperatures

Responses of pregnant females to a change in ration level

There is reason to suppose, but only little hard evidence, that a generous ration after weaning and in the first few weeks of pregnancy might enhance the number of ova released and the number of embryos safely implanted into the uterine wall. If there be any such response, it would be difficult to demonstrate on account of the multiplicity of factors

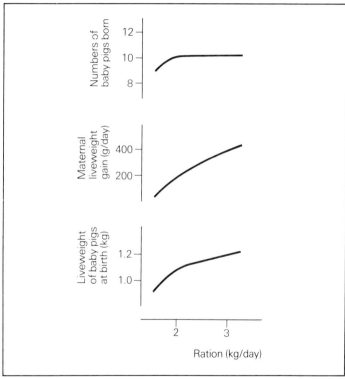

Figure 5.15 Responses of pregnant females to a change in feed allowance

influencing fertility, and because the female is by nature a variable beast. Body condition would predictably have much to do with it, the influence of feed level being greater with females in low condition. In the UK a ration of about 2.5–3.5 kg is given to newly weaned females and this may be continued after mating for some time into pregnancy. Animals in poor body condition could justifiably be rationed for short periods to a level of 4 kg daily.

Feed allowance has a more clear-cut effect on pregnancy itself (Fig. 5.15). At feeding levels below 2 kg, it is possible that the maternal body would need to be used up for purposes of supporting foetal gain. The offspring of shrinking dams are fewer and of lower birthweight than those carried under more normal circumstances. At feeding levels above about 2 kg, there is little influence of ration on numbers born, but a predictably positive association with maternal and foetal gain. As a general rule, over-fat females have wasted food, will have fewer young and are likely to have problems at parturition. Over-thin females produce weak small pigs, give less milk, are less likely to rebreed within five days of weaning and are more likely to suffer embryo losses.

The effect of ration upon birthweight is quite small, about 0.1 kg on the weight of each baby pig in response to an extra 1 kg of feed daily throughout the 115 days of pregnancy – a feed : gain ratio of about 100 : 1. Nevertheless pig birthweight begins to have significance when pigs of low weight die simply because they are small, as often happens to those of less than 1 kilo.

As thin sows do not readily rebreed, an increase of feed allowance for pregnancy is normally indicated when sow condition suggests the likelihood of there being insufficient fat stores to support the next lactation; and when the lactation has ended, to encourage rapid re-impregnation. With over-fat sows the appropriate action would be for a ration decrease (Fig. 5.16). The relationship between amount fed and body condition of the female is nebulous. Two-and-a-half kilograms of feed might barely support maternal weight stasis on some units, while on others the same ration would be grossly overindulgent.

In an attempt to achieve some degree of uniformity, it has been suggested that ration recommendations for pregnant females should be based on body condition as an indicator of the level of fat stores (Fig. 5.17). Or alternatively, as it is not possible to maintain body size and body condition concurrently, on the amount of maternal weight gained over each reproductive cycle (Fig. 5.18). Breeding females, particularly those in their first and second parities, are still growing toward their mature size, and if the ration does not aim to provide for some body gain, pigs will simply become emaciated.

Both body condition and weight gain methods are of greater merit than any sort of generalized recommendation (such as that all pregnant sows need a ration of x kg), but they are themselves imperfect. The amount of fat, or body condition, required to be on the back of a female has yet to be elucidated, and a reliable on-farm method of assessing fat stores is not yet with us. Weight gain as the criterion of an adequate ration has its proponents, and 15 kilo gain from one weaning to the next until the fourth parity (Fig. 5.18) has been suggested as a guide. But 15 kg can mean different things to the fat stores of female types ranging from lank to dumpy, and from 100 to 200 kg liveweight.

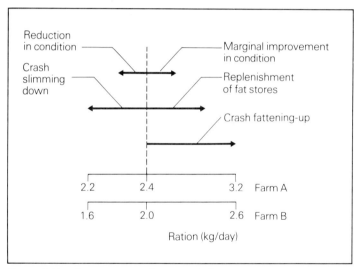

Figure 5.16 Consequences of a change in feed allowance to pregnant females on two different farms

Responses of lactating females to a change in ration level

Lactating pigs, like other lactating mammals, are designed to milk off their backs. Weight loss, particularly fat loss, is normal. Catabolized body fat may be considered a *bona fide* energy source for milk synthesis. The rate of fat breakdown to support lactation is commonly about 0.25 kg daily; but may be much greater, depending upon the ration supplied (Fig. 5.19). The greater the number of sucking pigs and the less the food intake the faster the stores of fat will be used up (Fig. 5.20). It follows that the responses of lactating females to ration changes must be largely a function of both the *number* of young in the litter and the *extent* of the body fat stores.

There are two areas of concern in rationing lactating animals; the body condition of the mother in relation to re-breeding, and the growth rate of the young. In the latter respect a further complication arises from young pigs having the possibility of two food sources; mother's milk and supplementary solid feed. While depending on milk for the first two weeks or so of life, the level of milk supply becomes increasingly less important after 14 days. It is not surprising that generalizations about the effects of ration allowance on performance of lactating pigs are likely to be of little practical use.

It does appear logical to feed at least to litter size (Table 5.1). But if pregnancy feeding is sufficiently expertly

Figure 5.17   (a)–(d) Examples of ration recommendations on the basis of body condition

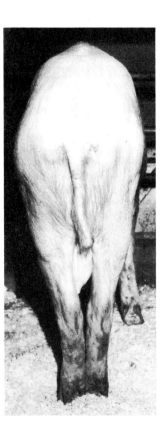

Figure 5.17   (a) Much too thin: feed 0.4 kg more

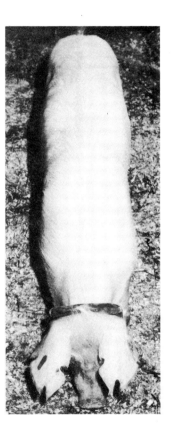

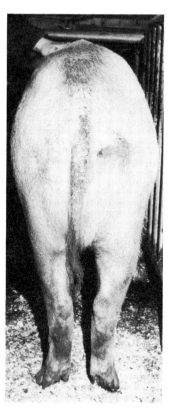

Figure 5.17 (b) Too thin: feed 0.2 kg more

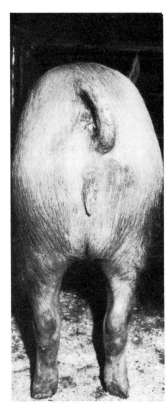

Figure 5.17 (c) About right

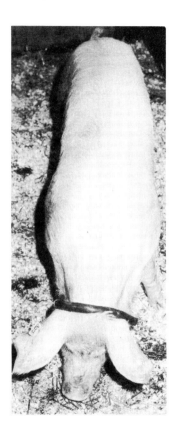

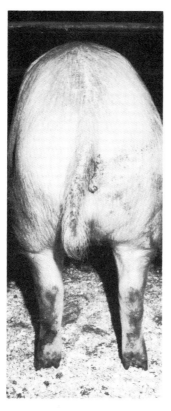

Figure 5.17 (d) Too fat: feed 0.2 kg less

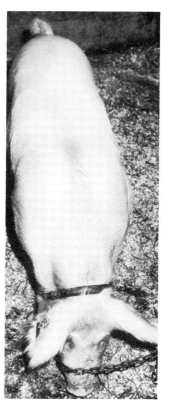

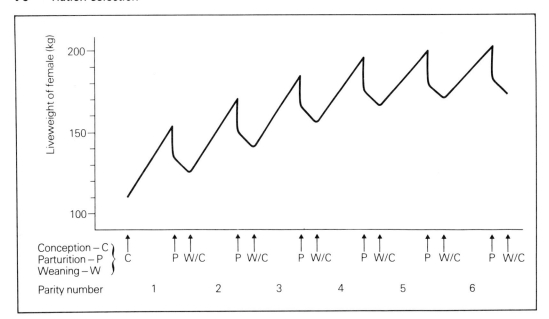

Figure 5.18 Pattern of liveweight change through breeding life of female pigs. In the programme shown the pig gains 45 kg each pregnancy, loses 20 kg at parturition and 10 kg during lactation. Weight gain from parity to parity is 15 kg until the 4th parity. Liveweight stabilises at around 180 kg

controlled to effect all the necessary adjustments to the female's body condition, then a flat rate for all sows of 5.5 or 6.0 kg daily might be justified. Under any flat rate feeding system however, there is an increased likelihood of both overfat and overthin animals arising within the herd.

Until the relationships between nutrition and reproductive performance have been resolved, feed allowance recommendations for pregnancy and lactation should rightly remain vague. The extremes of obesity and emaciation are simple to detect and their consequences self evident. But even

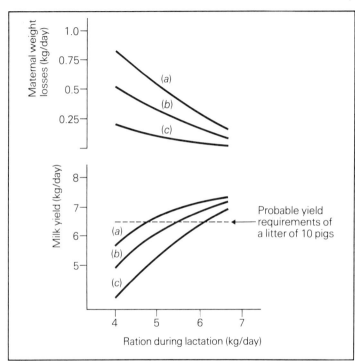

Table 5.1 Ration allowances for lactation

Litter size (number of pigs)	Ration (kg/day)
7 or less	5.0
8–9	5.5
10–11	6.0
12 or more	6.5

between these extremes, the range of rations which are 'satisfactory', as judged by performance, is still remarkably wide (1.8–3.0 kg in pregnancy, 4–7 kg in lactation, 2–4 kg between weaning and confirmed conception; and these ranges amongst 'standard' intensive conditions on 'normal' diets). The best basis for a decision on whether a change in ration allowance is needed is the evidence from the individual unit concerned. Figure 5.21 is given as an example of an appropriate diagnostic routine; rationing the breeding herd correctly will always require a flexible, not a stereotyped, approach.

Figure 5.19 Expected responses of breeding females to a change in feed allowance during lactation: (*a*) females in a good condition with ample body stores of fat available for milk synthesis; (*b*) females in satisfactory condition; (*c*) females in poor condition with low reserves of body fat

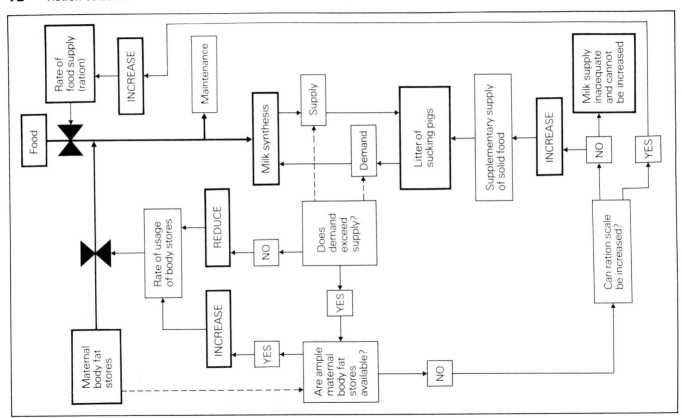

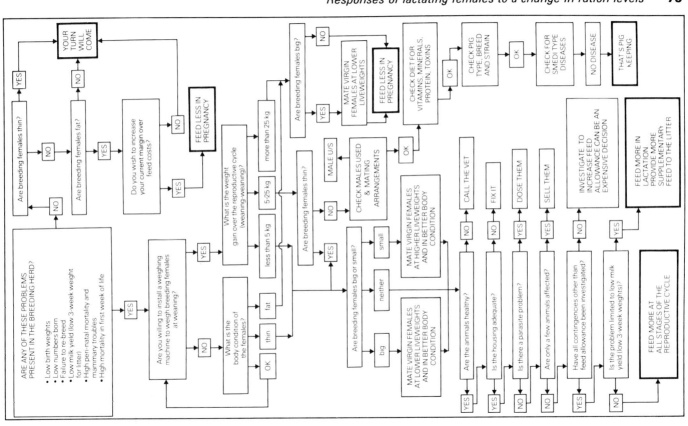

Figure 5.21 Diagnostic and action chart to assess adequacy of feed allowance to breeding female pigs

Strategy for meat production

6

The pig producer is at the centre of a web of interacting forces, all of which conspire at best to frustrate and confuse him and at worst to force him out of business. But the astute manager will endeavour to manipulate his circumstances rather than be manipulated by them. For this, purposeful strategic decisions are required. These need knowledge of the way the various forces act upon a system, and help is also required with the mathematics. Both can come from modern computer technology. It is impossible for the human brain to simultaneously contain the complexities of all the feasible courses of action and their consequences, never mind determining any one action as better than any other. Pigs can be fed a multiplicity of diets, of different prices, at innumerable different feed allowances, for various lengths of time, to slaughter at any number of weights between 55 and 120 kg. Consequent upon diet and feed level also comes carcass quality, which can determine the value of the end product. At each of all the possible slaughter weights there are a plethora of grading schemes, severe and lax, offered by a variety of market outlets wishing to put out contracts to producers for their pigs. If this were not enough, pigs are subject to a fickle market which causes both the costs of weaners and the values of slaughter pigs to fluctuate. The difference between prices paid for pigs of differing quality (the grade differentials or premiums) can also be altered at the whim of the buyer, as can the contract which states the requisite depth of backfat for each level of grade premium. Finally, add to all this the violent oscillations of the world feedstuffs market which controls the price for pig feed, and one begins to understand why access to a computer is as essential a component in pig production strategy as are the buildings and stock.

Because feed is by far the biggest part of pig production costs, strategic feeding is the door opening to maximization of profit margins. The key to strategic feeding is the ability to predict quantitatively pig response to a change in feed dose. A method for dose/response prediction lies at the core of a computer model developed at the University of Edinburgh in the 1970s, and now generally available. One of the uses of the Edinburgh Model Pig is to look at aspects of production strategy and financial optimization.

The criterion for profitability is always a moot point. Absolute profitability depends on knowing the fixed costs, and the gross margin depends on knowing the variable costs. Both these are highly individual characteristics of a farm. The important cost elements are those of feed and weaners. Perhaps the most useful measure of profit is therefore the margin between the price received for the slaughtered pig and the costs incurred to produce (or buy) the weaner, and to feed it. Margin over feed and weaner costs can be expressed on the basis of an individual pig (per pig), or in the basis of pen space available in a year (per pig place per year). The latter figure has the advantage of allowing for the fact that the real rewards of quicker growth, or of selling pigs at lighter weights, come from increased pig throughput. The example in Fig. 6.1 shows that although the best margin per pig is achieved when selling at a slaughter weight of 115 kg, best margin per pig place per year occurs at a slaughter weight of around 95 kg. As pig producers run farms rather than individual pigs, and spend money day by day rather than only when they sell a pig, the per pig place per year criterion is by far the more useful indicator of pig production profitability.

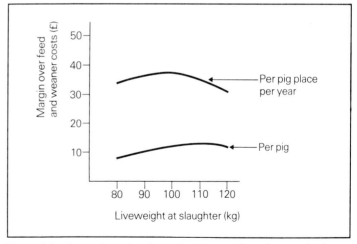

Figure 6.1 Comparison of profit margins expressed per pig or per pig place per year

When to slaughter

Figure 6.1 has already shown that bacon production can have the edge over other slaughter weights. This example was derived from the Edinburgh Model Pig set up as in Table 6.1 but the position is not clarified until the interaction with feeding level is analysed (Fig. 6.2). It is apparent that to exploit the grading situation within the bacon scheme (which, at a maximum of 16 mm P_2 for top grade, penalizes fat quite severely), low levels of feeding are much preferable to high levels. Indeed, at high levels of feeding there is little to choose between production of pigs for the pork and bacon markets.

If meat packers wished to encourage the production either of pork or of heavy pigs, they would need to improve the

Table 6.1 Details of some of the variables used in setting up example runs of the Edinburgh Model Pig to analyse production strategy

1 *Diet* Energy, 13 MJ DE/kg. Protein, 140 g DCP(165 g CP)/kg.
Biological value of protein, 65. Crude fibre, 40 g CF/kg. Cost, £120/tonne.

2 *Pig type* Strain, improved. Cost, £25 for 20 kg liveweight 'weaner'.

3 *Housing* Quality, average – slightly draughty. Temperature, about 18 °C.
Floors, part solid and insulated (lying area), part slatted.

4 *Ration scales (feed allowances)*

	High	Medium	Low
For pigs at 20 kg liveweight:	1 kg	1 kg	1 kg
Raised weekly by increments of:	0.25 kg	0.20 kg	0.15 kg
To a maximum limit level of:	3.0 kg	2.6 kg	2.2 kg

5 *Carcass classification grade schemes*

	Backfat depth (mm P_2)	Price/kg deadweight (p)
Pork (55–65 kg deadweight; about 80 kg live)		
Grade 1.	<13	95
2.	13–16	90
3.	16–19	85
4.	>19	80
Bacon (65–75 kg deadweight; about 95 kg live)		
Grade 1.	<16	90
2.	16–19	85
3.	19–22	80
4.	>22	75
Cutter (75–85 kg deadweight; about 105 kg live)		
Grade 1.	<19	85
2.	19–22	80
3.	22–26	75
4.	>26	70
Heavy (85–95 kg deadweight; about 120 kg live)		
Grade 1.	<22	80
2.	22–26	75
3.	26–30	70
4.	>30	65

prices for these weight categories and/or reduce the price penalty discrimination against P_2 backfat thickness. A relaxation of grading standards would prompt in turn the use of higher levels of feeding. Such moves on the part of pig meat buyers should be made in cognisance of general market forecasts. For example, an increase in the price of raw materials for animal feed – particularly cereals – would of itself tend to bring about a reduction in weight at slaughter and a restraint upon liberal feed usage. In some countries it is common practice for flat-rate payments to be made for all pigs; this stimulates higher rates of feeding, heavier slaughter weights and fatter pigs.

The cost of producing, or buying, weaners has the classical effect of any unavoidable cost upon a production system; as the cost of weaners goes up then the heavier becomes the

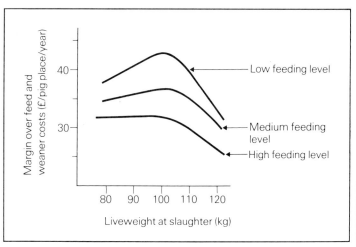

Figure 6.2 Profitability of pigs slaughtered at various weights after having received different feed allowances (see Table 6.1 for details)

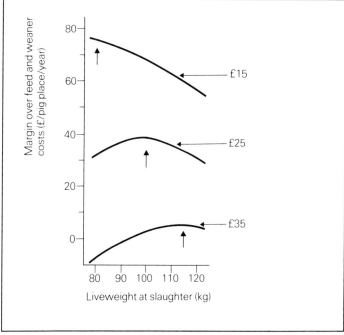

Figure 6.3 Influence of 20 kg weaner costs upon optimum slaughter weight

optimum slaughter weight (Fig. 6.3). At £15 apiece for weaners, marketing at pork weight is the requisite tactic; at £25, bacon; at £35, heavy.

How much to feed

The amount fed is related to most of the other controllable factors in pig production management, not least – as has been shown – with slaughter weight. Within any one weight, the appropriate feeding scale depends heavily upon the type of grade classification and the grade premiums offered. Table 6.2 details four alternative schemes. Scheme 1 has been used previously, while Schemes 2 and 3 are increasingly lax with regard to the amount of backfat allowed. Scheme 4 is as for Scheme 1 for backfat limits, but the grade premiums are less. The influence of these schemes upon the selection of optimum feed allowance is depicted in Fig. 6.4. The point at issue is not merely that the more lax are the requirements of the grading scheme the more profitable is pig production, but that different levels of feeding are needed to cash-in on the opportunities offered by different schemes. Thus for Scheme 1, low feed levels optimize; for Scheme 2, medium feed levels; for Scheme 3, high feed levels. In recent years in UK contracts offered to producers have progressively reduced the P_2 backfat depth maxima. To optimize their margins, producers would have required to have taken positive action appropriate to the new conditions prevailing.

Table 6.2 Alternative schemes for grading and pricing bacon pigs

	P_2 backfat depth (mm)	Price per kg deadweight (p)
Scheme 1		
	<16	90
	16–19	85
	19–22	80
	> 22	75
Scheme 2		
	<20	90
	20–23	85
	23–26	80
	> 26	75
Scheme 3		
	<24	90
	24–27	85
	27–30	80
	> 30	75
Scheme 4		
	<16	90
	16–19	88
	19–22	86
	> 22	84

Scheme 4, which offers reduced quality premiums, yields greater margins at the higher feed levels, thus opening up the benefits of producing faster growing, lower quality pigs. The difference between the margin yielded at the lowest and the highest levels of feeding was however only £3.

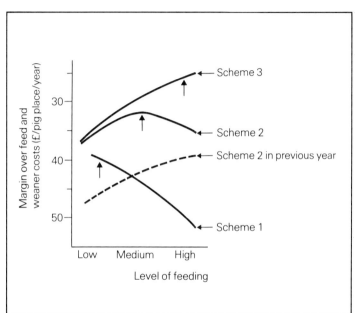

Figure 6.4 Relative profitability of different schemes for grading and pricing bacon pigs (Table 6.2), and the relationship between scheme in force and optimum feeding level (Feed levels as in Table 6.1). It is an important principle of these relationships that they will differ as the circumstances of feed and pig prices change. Scheme 2 may in one year be best exploited by a low level of feeding, while in another year by a high level of feeding. Which level is appropriate must therefore be calculated freshly each time the situation changes

What to do about sex and strain

Pigs of improved genetic strain – and sex in the order castrated male, female, entire male – will respond positively to increased feeding level (Fig. 6.5). The production of entire males is greatly the more profitable: castration, at a stroke, converting the best pigs into the worst. The poorer lean potential of the castrate results in margin maximization occurring at the lowest feeding level, while for females the medium feeding level is

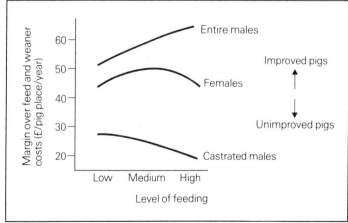

Figure 6.5 Influence of pig type upon profitability and choice of feed level (Grade scheme 2 (Table 6.2), other details as Table 6.1)

optimum and for entire males the highest feeding level. In general, the producer will enhance the benefits of better pigs by commensurately more liberal feeding. Actual feed levels employed will naturally interrelate with the grade scheme; the more severely the scheme discriminates against fat, the less the benefits of liberal feeding and the more important the use of improved pig types.

House temperature

Effective house temperature can be lowered not merely by a fall in air temperature but also, for example, by an increase in air flow through the house and the loss of heat through the floor. As cold pigs use up feed to keep themselves warm, both growth rate and feed conversion are adversely affected, with consequent losses to potential profit margins (Fig. 6.6). Generous feeding can go some way towards ameliorating the effects of cold.

Choice of diets

Pig diets can be prepared to a range of specifications each with a different cost. Three diets were tested through Edinburgh Model Pig and the results were as shown in Table 6.3. Costs for the diets appear to have realistically reflected their worth,

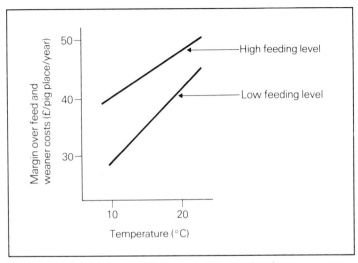

Figure 6.6 Profitability of light pork production at various house temperatures

thus with pigs produced for Grade Scheme 1, low feeding levels of any of the three diets would have resulted in similar profits. In the case of pigs destined for Grade Scheme 2, diet A, although the cheapest, was over-priced, while low feeding levels of diet C and medium feeding levels of diet B yielded the best margins. Buyers of diet A would have needed to feed to the medium or high ration scales to make optimum use of it.

Table 6.3 Margins over feed and weaner costs (£/pig place/year) for three diets (A, B and C) of differing quality and price offered at three feeding levels to pigs destined for either of two bacon grade schemes

Feed allowance[b]	Grade scheme 1[a]			Grade scheme 2[a]		
	Diet A[c]	Diet B[d]	Diet C[e]	Diet A	Diet B	Diet C
Low	40	41	40	41	45	48
Medium	37	36	34	46	48	47
High	33	31	31	46	45	45

[a]See Table 6.2.
[b]See Table 6.1.
[c]Diet A – 12 MJ DE/kg, 130 g DCP(153 g CP)/kg, £112.5/tonne.
[d]Diet B – 13 MJ DE/kg, 140 g DCP(165 g CP)/kg, £120/tonne.
[e]Diet C – 14 MJ DE/kg, 150 g DCP(176 g CP)/kg, £127.5/tonne.

Table 6.3 also provides an object lesson in the impossibility of predicting the outcome of one set of circumstances by analogy with results from another. Thus a judgement as to which diet was best under conditions of high level feeding and Scheme 1 would be useless for a situation involving low level feeding and Scheme 2.

Use of models

For the planning of practical feeding strategies prediction models should be used semi-continuously. So many of the interacting forces can change in so many directions that the computer must be set up afresh with each twist and turn of circumstance. Examples of results from models can only be used to *illuminate* some of the profit margin responses to management changes. Each response will differ between, and depend upon, a multiplicity of interacting circumstances – only a small sample of which have been examined here. Such examples of particular events cannot be taken as generalizations; the temptation to derive general rules from specific findings has bedevilled the useful application of agricultural research into farming practice for long enough. Strategic planning on the production unit requires frequent use of the models themselves at first hand, not interpretation of examples derived from them.

Pig breeding

7

However good he may be, a producer cannot do better than is within the capability of his pigs, and the limits of pig performance are controlled by the animals' genetic make-up. Improvement in pig performances can thus be achieved by selection of animals with superior genetic constitutions as the parents for the next generation.

Pig breeds

Breeds of pigs in Europe may be classified into two broad groups: white pigs and exotics. There are three principal white breeds: Large White (Yorkshire), Landrace and Welsh. Differences between the white breeds are not readily apparent visually, except for the position of the ears: respectively up, down and down. The Large White is slightly more prolific, faster growing and the meat is of better quality. The Landrace is mostly used for crossing, the Large White × Landrace pig performing even better than the pure-bred Large White. About 75 per cent of pigs in the UK are crossbred. The British Landrace is of Scandinavian origin, while in addition to Swedish, Norwegian and Danish Landrace breeds there are also Belgian and Dutch varieties. Ham shape can differ between breeds and strains, and this can be important for some markets.

Exotic breeds include the Pietrain (from Belgium), and the Hampshire, Duroc, Poland-China and Lacombe (from North

America). Many of the exotics are coloured to a greater or lesser extent, have shorter carcasses with bigger eye muscles and chunkier hams and have poorer reproductive performance. The old English breeds such as the Saddlebacks, Tamworth, Large Black and Gloucester Old Spot are now largely fancy breeds often kept on outdoor extensive systems.

Limits to production

One measure of the efficiency and profitability of a pig enterprise is the ratio of *food costs* to *animal value* traded in the year. The characteristics primarily involved are the daily rate of lean growth and the propensity of the animal to lay down fat in the carcass. Connected are factors such as appetite (which, if low, holds back lean growth), length (for any given mass of fat, the shorter pig will have a deeper layer of backfat), shape (which may impart extra market value) and the quality of the lean and the fat in the meat. The growing pig would therefore be improved by possessing characteristics of faster lean growth and lower carcass fatness, together with adequate appetite and meat quality.

In the breeding herd the universal index of production success is the number of piglets weaned per breeding female each year. The major controlling force of this index is the number of piglets born, with some additional provisos about the female's readiness to rebreed at the end of the lactation and the willingness of young pigs to stay alive after being born. The improved breeding female is therefore one who produces numerous young at each confinement, who exhibits an unrelenting desire to be pregnant and whose offspring thrive.

Number of breeding objectives

There are some threshold characters that every breeding animal must have, regardless of the ultimate objectives of the breeding scheme. Leanness is only beneficial if the muscle is neither too dark nor too pale, neither too dry nor too wet. However excellent the growth rate, a male is no use unless standing on four good legs with an affinity for the opposite sex. A sow's ability to produce piglets is irrelevant if there are too few nipples.

Given that the threshold tests are passed, most rapid improvement will be made if selection is on the basis of a single objective. Rarely however is only one character in need of improvement and single-mindedness can neglect other important breeding objectives. Nevertheless, the more criteria that are included in the selection of an individual, the slower is the progress in any one direction. Then there is the problem of the relative importance, or weight, to be attached to each character, which requires the construction of an index. A simple index for two characters might be, $I = (0.08 \times$ daily gain$) + (50 -$ backfat depth$)$. Two males growing at 800 and

900 g daily with ultrasonic P_2 measurements of 15 and 20 would score respectively 99 and 102, so the faster growing but fatter animal is chosen. If carcass grade was considered more important than growth rate, the index could be weighted accordingly: for example, I = (0.04 × daily gain) + (100 − 2 × backfat depth). The pigs would now score 102 and 96, so the slower growing but less fat male is chosen.

Variation and inheritance

The ultimate form of the character to be selected must be available amongst the genes present in the existing pig population, which is to say that one cannot have as an objective a quality which may not exist. The task of the breed improver is to identify those individuals carrying genes capable of improving the character required and to combine them into a nucleus of improved stock. The average performance is raised by dissemination of the improvement through the general population.

The variation shown by a characteristic within a population of pigs will follow a distribution curve such as in Fig. 7.1A. The figure refers to P_2 backfat depth, which is shown in the original population to vary over the range 32 mm to 8 mm. A few pigs will be of a character similar to the ideal desired (P_2 about 12 mm), most will be average (P_2 about 20 mm) and a few will be of a character opposite to that desired (P_2 about

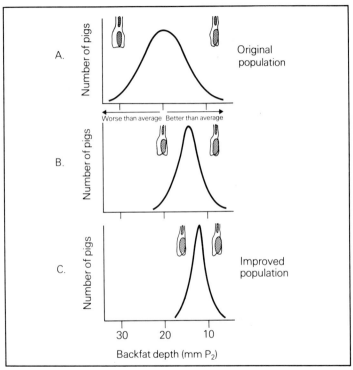

Figure 7.1   Distribution of backfat depth in a population selected for low fat. The symbol 🥓 represents a bacon rasher

30 mm). By picking out as parents for the next generation animals with characteristics closest to the ideal, the population average will move in a positive direction (Fig. 7.1B). The 'better' the chosen parents are, the fewer they will be, but the faster will be the rate of improvement. The population average closes up with the performance of the best animals and the variation diminishes. The less the variation in a population the more difficult it is to identify the better animals and the less is their effect upon the population average (Fig. 7.1C).

Inheritance is not an all-or-nothing affair; there is a gradation between zero and unity which relates to the likelihood of offspring inheriting a particular character

possessed by the parent. With pigs, reproductive qualities are of low heritability. The reason for one female producing more young than another is most likely due to environmental circumstances and little to do with the genetic make-up of the animal. Factors which control growth seem to be more highly heritable, so improving the chance of better pigs producing offspring which are also better. Estimates of heritability are shown in Table 7.1.

Progress in pig improvement is related to the heritability, the rapidity with which the next (improved) generation comes along and the size of the difference which the breeder can achieve between selected parents and the average of the population – that is, how much 'better' selected animals really are (Fig. 7.2).

Where variation in a character is limited in one population of animals, then another population might have something to offer. The simplest way to increase the length of Hampshire pigs (average length at bacon weight 750 mm) is to crossbreed them with the Large White (average length 800 mm). By the same token, a chunky shape is best imparted to a Large White population, not by selecting chunky pigs out of that population – always supposing that there were chunky Large Whites to select – but by using a parent which contains some Pietrain, Hampshire or perhaps even Belgian Landrace blood.

Table 7.1 Some estimates of heritability (coefficient)

Reproductive characters

Numbers born	0.1 –0.2	
Survivability of young	0.05–0.1	Lowly heritable
Readiness to rebreed	0.05–0.1	
Milk yield	0.1 –0.2	

Growth characters

Lean tissue growth rate	0.3 –0.5	
Backfat depth	0.4 –0.6	
Carcass length	0.4 –0.6	Moderately heritable
Eyemuscle area	0.3 –0.5	
Meat quality	0.2 –0.4	
Appetite	0.4 –0.6	

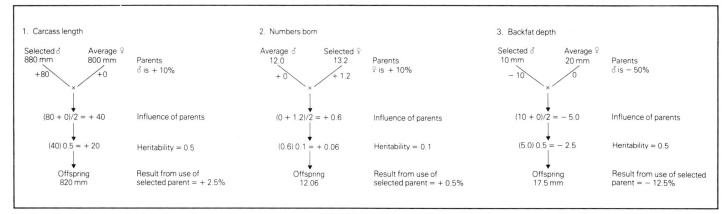

Figure 7.2   Influence of heritability and amount of variation upon rate of improvement following use of a selected parent. In the first example, the selected male has a carcass length 10% better than the average breeding herd. If the heritability for carcass lenght is 0.5, then a 2.5% improvement would be found in the offspring. Some 12 months are likely to elapse between the birth of the selected male and the birth of his offspring. In a less variable population of pigs a male with a carcass lenght of 880 mm might not exist, and the longest animal be only 5% above average (840 mm), so the offspring would only be improved by 1.25%.

In the second example, a breeding female has been selected consequent upon her producing litters with 10% higher than average numbers of pigs. With a heritability of 0.1, the offspring are likely to show an improvement of only about 0.5%. In this case, 24 months or more elapse between the birth of the selected female and the birth of her offspring.

In the third example a male with P_2 10 mm, used on a population with average P_2 20 mm, will produce offspring whose P_2 measurement will be about 17.5 mm (heritability assumed to be 0.5).

A male with a P_2 measurement of 14 mm would, by the same calculation as Example 3 produce offspring P_2 measurements of around 18.5 mm. The size of difference between the average of the selected parent and the average of the population will depend upon the number of animals which have to be selected from amongst the candidates. It is, for example, much more probable that the mean P_2 of those selected will be 10 mm if one is selecting only 5 individuals from 100 candidates; and conversely, more probable that the mean P_2 will be 14 mm if one were selecting 25 individuals from 100 candidates (Figure 7.3)

Hybrid vigour

Hybrid vigour describes a phenomenon which brings about performance levels greater than would have been expected from an animal's genetic constitution. With pigs, reproductive performance is improved when some specific animal types are crossed. One of the best documented cases is when the Large White breed is crossed with Landrace. The number of pigs born is increased such that crossbred litters can be 5 per cent larger and 10 per cent greater in total weight than purebred litters. If the mother is also herself a crossbred, then the improvement over the purebred is even greater, due to enhanced mothering qualities (Table 7.2). Crossbred dams also show a 10 to 15 per cent reduction in weaning to conception interval. Hybrid vigour is exploited simply by using crossbred, rather than purebred, animals as the dams of the slaughter generation, and choosing for the sire an unrelated breed or strain. Hybrid vigour is not a substitute for breed improvement by selection (merely 'adding value' to any such improvement) and there is no reason to slacken off

Table 7.2 Effect of cross-breeding upon reproductive performance

	Percentage improvement over purebred for pigs weaned/dam/year
First cross	5–10
Back-cross or three-way cross	10–15

endeavour to increase reproductive performance of grandparent purebred stock by conventional means.

The test environment for growing pigs

To help select pigs of high genetic merit, candidate animals can be given a special test regime.

The simplest method of feeding and penning animals on test is to house in groups, feed *ad libitum* and select on an index of growth rate and carcass quality. This regime has tended in the past to lead to a dramatic reduction in backfat depth and some reduction in appetite. These would be ideal responses if the commercial environment is one of *ad lib* feeding, but would not necessarily help to improve daily lean growth rate. In the UK P_2 backfat measurements of below 12 mm are probably undesirable and there is not much further improvement to be made in the better strains of pigs, particularly as backfat can be controlled by judicious rationing and leaving males entire.

Individual penning of candidates allows an increase in the precision of test. A popular regime is to place the pigs on test at a given weight (preferably 20–30 kg, but more often 30–40 kg), and then to feed according to a fixed ration scale which increases on a weekly basis. The test ends after a predetermined number of weeks (often 12), and the animals are selected on liveweight and backfat measured ultrasonically.

As all the pigs have been given the same amount of food, it is probable that the heaviest animal at the end of test will have had the fastest lean growth. The measurement of fat depth is

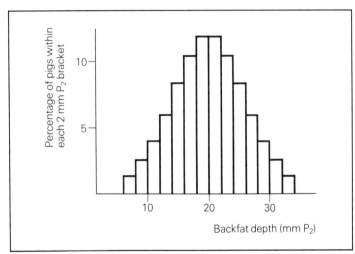

Figure 7.3 Example of the variation in P_2 measurements taken from a population of pigs out of which males are to be selected.

If the number of candidates is 100 and 5 require to be selected, then the highest P_2 measurement in the selected pigs will be about 11 mm and the average about 10 mm.

If the number of candidates is 100 and 25 require to be selected, then the highest P_2 measurement in the selected pigs will be about 16 mm and the average about 14 mm.

of interest in its own right and also gives collaborative evidence to the estimation of lean growth.

The dilemma with the fixed feed/fixed time trial is in the choice of ration scale. If the scale is too restrictive, the best pigs will not maximize their lean growth and there will be bunching of performances, making identification of genetically superior animals more difficult, although it is possible that in these circumstances there is pressure favouring the selection of animals which discriminate against fat – 'thinnies' (see Fig. 3.18). If the scale is too liberal, a proportion of the candidates will not eat all the food, which would favour individuals with good appetites. The difference between *ad lib* feeding and feeding to a scale so liberal that a proportion of the pigs cannot eat it, is a matter of degree. The response that is likely to occur is much dependent upon the weighting given to fatness in the selection index.

There is the alternative possibility of feeding candidates for selection according to their liveweight. This seems logical as it provides more food to the fast growers and increases the spread of candidates. But it is a difficult regime to implement, requiring frequent weighing of pigs, as well as suffering from the drawback that pigs slow to start have no means of redeeming themselves and are restricted throughout, regardless of merit.

The end of the test also poses its problems. Finishing at an exact predetermined weight, spreads liveweight gain and fatness between individual pigs, but there is a complex interaction with feed use. To terminate the test at a predetermined time is tidy, but the spread of fatness and growth rate may be reduced.

Currently two test regimes seem in favour. One is *ad lib* feed to a fixed slaughter weight and selection on growth rate, backfat depth and feed efficiency. The likely response is a decrease in both appetite and backfat depth, with perhaps some small improvement in lean growth. The other is fixed feed according to week on test, fixed slaughter time and selection on lean growth rate as estimated from liveweight and fatness. Depending on the proportion of the pigs refusing feed, the hoped-for response is an increase in both appetite and daily lean growth rate.

The pedigree (nucleus) breeder

Improving pigs by genetic selection implies generation by generation progress in a population of foundation (nucleus) stock. To do this a breeder will concentrate on one pure breed and usually only on one, or perhaps two, strains within that breed, taking the best individuals as the parents of the next generation. A strain might be selected for one character, perhaps lean growth rate, or for a complex of growth characters by use of an index, together with positive selection for reproductive performance. Growth characters may be selected for in one strain, and reproductive performance in

another, thereby producing two specialized lines.

The pedigree breeder is always on the lookout for really outstanding individuals – males with very rapid growth and females with extreme prolificacy – and once he has found them will tend to hold them in the herd for as long as possible, or make his reputation by their sale. If progress slows, or a neglected character is slipping back, it may be opportune to import new genes into the herd in the form of stock from another breeder. The availability of semen from males of merit standing at artificial insemination centres makes this a relatively simple move.

It is difficult to make progress in reproduction performance, both because of the number of litters needed for a breeding female to prove herself as being of merit, and the likelihood of circumstance confusing the issue. Many breeders resort to a policy of merely ensuring that all candidates for selection are from large litters, which is unlikely to improve reproductive performance, but may prevent its decline.

Growth characteristics are more easily tackled, and can be measured directly in both males and females. As the proportion of female candidates necessarily selected is high (about 20 times higher than for males), then accurate testing is not particularly cost beneficial. Candidates for selection as breeding mothers may be taken from group fattening pens at 80 kg, their growth rate determined from their age and the backfat depth estimated with the use of an ultrasonic meter. Liberal feeding of candidate pens would ensure that no animal

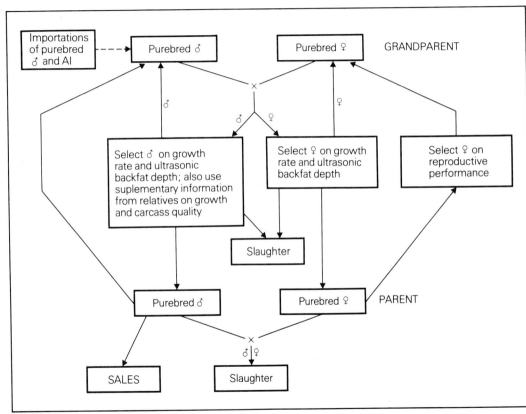

Figure 7.4 Example of breeding scheme for pedigree (nucleus) breeder

is prevented from demonstrating its potential because of an inadequate food supply. But where carcass quality is a significant factor in the value of slaughtered animals, a large number of fattening pens put on to a liberal feeding regime could be costly, as females not selected are likely to be downgraded as over-fat.

There is more reward for increasing the accuracy of tests for male candidates. A greater differential between selected animals and herd average is possible as the numbers required are few, and the influence of the male on the herd is deeper and longer lasting. Candidate males can be penned separately, or in small groups of two or three brothers, so that some estimate of intakes and feed efficiency can be made. In addition to growth rate, feed efficiency, and ultrasonic backfat measurement, it is helpful if information is brought to bear from relatives, especially about carcass quality of less fortunate slaughtered brothers and sisters.

Central testing can offer help to pedigree breeders by providing facilities at which candidate males and their relatives can be grown and assessed. To test for carcass quality, slaughtered relatives can be dissected into lean, fat and bone, and the muscle quality assessed. At central testing stations the breeder's stock can be scored in comparison to contemporary stock on the same test. On the basis of the score achieved the breeder can decide whether the animal should be returned to his herd or sold off. There can be a reciprocal arrangement whereby the central testing agency publish the performance characteristics of animals tested. The information is of benefit to other breeders, and enables pigs of merit to remain in herds of merit.

The breeding company

The companies have the advantage of size. With upwards of 1000 sows, the benefits of picking a few animals from a large diverse population can be exploited, and there can be a wide difference between the best selected great-grandparent males and the average of the pig population upon which they are to be used. The breeding company is not looking for a few animals of outstanding merit so much as for a large number of animals that are improved to some extent. Progress is made by frequent small steps forward, and the turn-round time in such a scheme is crucial. Stock sires are not kept for as long as possible, but for as short as possible, so the next generation contributes its small effort to positive progress as soon as it is able.

Whereas the pedigree breeder stakes his reputation upon the sale of a relatively small number of males of outstanding merit, the breeding company is in the animal supply business. Breeding companies sell both males and females, often in the form of a stock package. The purchaser is offered brand-named parent females as herd replacements, together with parent males specifically bred as sires for use with the branded females. Breeding companies will also offer – at higher prices – grandparent stock to those wishing to breed their own replacements.

The company therefore provides two services: trouble-free provision of male and female herd replacements, and improved pig stock. But it is only by the latter that the company will stay in business. It is fundamental to commercial success in the open market that the goods offered are demonstrably competitive in terms of quality and price: to be a step ahead, with stock of improved growth rate, carcass quality and reproductive characteristics. Breeding companies will test and select in their purebred nucleus herds in a similar way to pedigree breeders, and in addition will cash in on hybrid vigour by presenting a crossbred female as the branded product.

Besides considerations of improved genetic merit, companies concerned with the sale and movement of stock are rightly preoccupied with the health status of the animals. Disease prevention and control is a significant part of company policy.

The structure of the breeding company reflects its needs. Nucleus (great-grandparent) purebreeding herds are maintained in strict isolation. Within these herds vigorous testing for growth, carcass quality and reproductive performance is undertaken on both male and female lines. Selected out of these nucleus herds are the grandparent stock, a few of the best of which are used to refurbish the nucleus,

while most others are sent to the multiplier herds which generate parent stock for sale. Because of health control, animals once shipped to multipliers cannot come back into the nucleus. To improve reproductive performance in the nucleus herds, proven females may be contract mated to the best males and the offspring earmarked for selection back into the nucleus rather than going on to the multiplier phase.

The type of purebred grandparent maintained in the nucleus herds has developed from three requirements. First, stock must already be of the high genetic merit, but open to further improvement. Next, the two types chosen for the nucleus herds must, when crossed out, exhibit hybrid vigour. Last, the breeding company may wish to maintain a nucleus herd of a variety of breeds of pigs which might impart, currently or in the future, special characteristics into the final product.

A breeding company may maintain, for example, two large nucleus herds, one of Large White type and the other of Landrace type. The pure breeding herds may number 100 to 500 sows or so. Smaller 20 to 40-sow purebreeding colonies of breeds of pigs such as the Hampshire, Pietrain, Duroc, Belgian Landrace, Lacombe or Saddleback may also be maintained in the hope of their being found useful as providers of characteristics such as hardiness, chunky shape and lean mass. Alternatively, or in addition, a line may have been developed which contains an amalgam of various breeds and this is then treated as if it were a new purebreed. In either

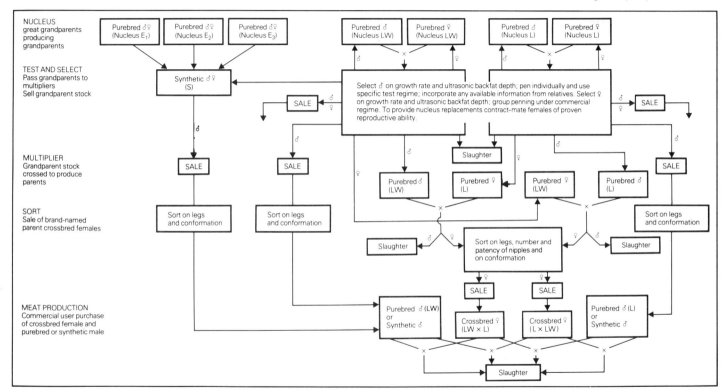

Figure 7.5 Example of a breeding scheme for a breeding company. E₁, E₂ and E₃ represent three different exotic breeds. S represents a synthetic 'sire line' made from the three exotics and the main breed type. LW and L represent Large White and Landrace types. These breeds are to serve as examples

case, the result can be a new *synthetic* line particularly appropriate as a top crossing sire at the slaughter generation level.

Figure 7.5 gives an example of how the organization of a breeding company might be put together.

The products sold from breeding companies may be tested against each other in central testing facilities. The Central Testing agency (the MLC in the UK) purchase from multipliers random samples of young females and mate them with boars supplied from the company nucleus. The females are tested for reproductive performance, and their offspring for growth rate, feed efficiency and carcass quality.

The commercial breeder/grower

Some commercial growers, although concerned to improve the genetic merit of their herds, may not wish to buy ready-made crossbred parent-generation replacements from breeding companies. Improved grandparent males can be purchased from pedigree breeders, or breeding company nucleus herds, or in the form of semen from AI stations. Purebred parent males used by the grower to generate pigs for slaughter can also be purchased from outside – often as close relatives of males of proven excellence – or they may be home bred. In any event, to generate crossbred parent gilts, the breeder must maintain some purebred nucleus sows of at least one of the two required breeds.

Testing may often be limited to female selection for growth rate and ultrasonically measured backfat depth, together with rejection of animals on the basis of poor reproductive performance. Females found to be particularly excellent would be singled out for contract mating to refurbish the nucleus. In the top half of Fig. 7.6 the actions are those of a catholic pedigree breeder, while in the bottom half of the figure the actions are those of a multiplier. Some producers might opt to forgo the joys of purebreeding. For them, breeding companies offer grandparent stock of two breeds for multiplication on the farm. There would be little point in making a selection of the crossbred stock so produced, and the animals are sorted only on the basis of adequate legs and nipples.

The breeding scheme may be further simplified by use of artificial insemination, which can entirely replace the need for males at the grandparent nucleus and multiplier levels. Flexibility is also possible between the females in the grandparent nucleus and in the multiplier group, animals being interchangeable between these two levels. The nucleus female is essentially any purebred animal selected for contract mating.

The commercial grower

The object of the commercial grower is not to become involved in the breeding of improved pigs, but merely to

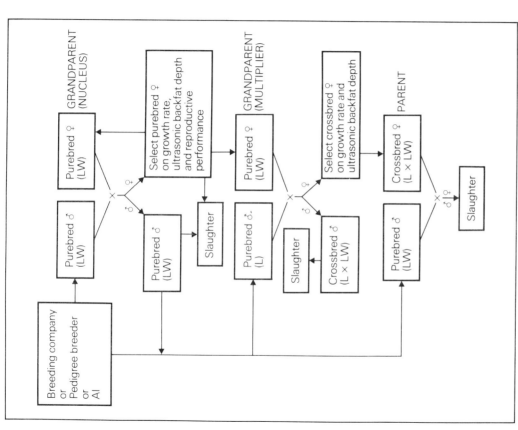

Figure 7.6 Example of breeding scheme for commercial breeder/grower. (LW and L represent Large White and Landrace breeds to serve as examples).

For each 100 crossbred parent females about 30 purebred animals are required. Of these about 10 regenerate the nucleus while 20 or so are mated with a male of a different breed to produce the crossbred females from which 50 replacements are selected yearly for the main herd.

The size of the purebred herd depends first upon the proportion of pigs rejected because poor legs, conformation and number of patent nipples, and second upon the proportion of pigs rejected as having failed to achieve adequate performance during the growing period. If 100 crossbred females have four litters each in the course of two years of reproductive life, then 50 replacements are required yearly. Twenty purebred females will produce about 320 piglets per year of which half will be female. If two-thirds to three-quarters of these animals are forwarded to test, then about half those candidates need to be selected for herd replacements.

To refurbish the nucleus at a rate of 15 females annually, 10 contract mated females will generate 80 female piglets, of which 50–60 would go forward to test and the top third of those candidates would need to be selected procure them. The breeding companies service him direct by the provision of crossbred females and selected males to go with them (Fig. 7.7), or indirectly by provision of grandparent stocks for farm multiplication of females (Fig. 7.8). Growers not wishing to depend on the breeding companies as a source of supply may maintain a crossbred herd by simple selection on the basis of leg and nipples, and hope for some

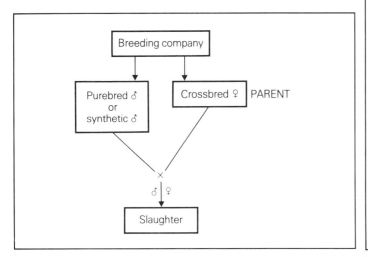

Figure 7.7 Example of breeding scheme for commercial grower purchasing parent stock from a breeding company

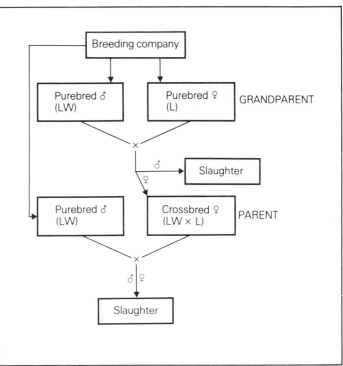

Figure 7.8 Example of breeding scheme for commercial grower purchasing grandparent (purebred) stock from a breeding company. (LW and L represent the Large White and Landrace breeds to serve as examples)

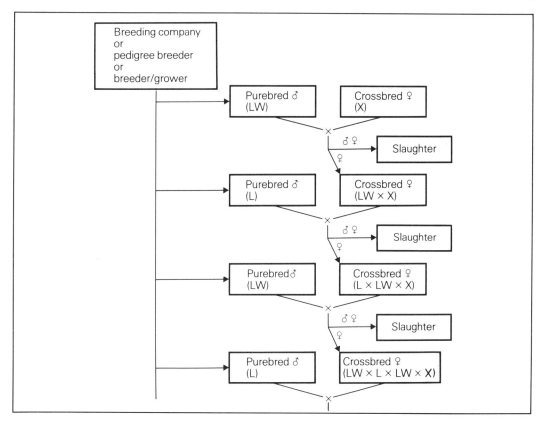

Figure 7.9 Example of breeding scheme for commercial grower

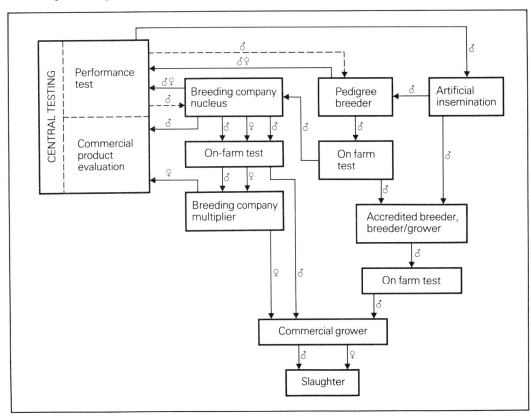

Figure 7.10 National structure for progressive pig improvement by identification, selection and dissemination of pigs of above average merit

improvement in genetic merit by the purchase of males from reputable herds (Fig. 7.9). To sustain the benefits of heterosis, the breed of male may be alternated every so often. Although purebred boars are the more likely option, some producers may go for a crossbred boar.

A national structure

Of the total population of pigs a very small proportion is in the hands of breeders; but the influence of the breeder upon the efficiency of both an individual commercial grower, and of the national herd in general can be far reaching. A national structure for pig improvement allows improved pigs to filter down from the purebreeding nucleus herds through the multipliers and into commercial slaughter pigs. Alongside, the central testing facilities – usually a Government or quasi-Government agency – help to identify the presence of genetic merit in purebred stocks and to compare the results of breeders' efforts.

Over the past 20 years of genetic improvement in the UK there has been such a marked reduction in backfat depth that there is little further benefit from continuing active selection against fatness. Progress in lean tissue growth rate and reproductive capacity has been slower, and it is likely that future selection pressure will concentrate upon these traits.

Housing

8

Buildings serve both to restrain pigs and to protect them from the elements. At different stages of their development pigs have differing environmental and physical needs, thus pig houses come in a wide variety of structural designs.

The principles of housing are simple, but in practice quarters for pigs often fail. The pig needs an area in which to lie warm and dry and an area in which to feed; it requires also to excrete, move about and have its being. The building must allow man to provide the feed, handle the stock and remove the excreta – preferably automatically. The whole to be accomplished at least cost. It is harmonizing the requirements of man and pig that is so difficult.

Confinement

Pigs kept under feral conditions exercise their free-will, but choose to live in self-imposed confinement. Groups will lie together in nests, and take shelter from the hostile elements of the weather under trees and scrub. A pig in a nest under a tree is 'confined'. A housed pig is more environmentally comfortable than a feral pig but, unless special provision is made, unlike the feral pig it cannot have any part of its life away from its fellows. In particular, this may result in disproportionate food sharing, physical damage through fighting, and a predisposition to rapid disease transmission.

Even when farmed outdoors, suckling females with their

young are often given individual field areas and invariably have individual housing facilities. One of the major campaign successes of the 1950s was to encourage the breaking up of mobs of pregnant females to allow the weaker ones a sporting chance of a fair share of feed. This in turn led to the provision of individual feeders for all adult breeding females.

Once under cover, pigs are confined for a variety of reasons, not least of which is to prevent damage to the fabric of the building and the pig's ultimate escape. To make most use of the money invested in enclosed areas within buildings, pigs should be kept as intensively as is feasible; but proximity causes one animal to trespass upon another's 'personal space', with ensuing loss of welfare and increased predisposition to disease. The implications for animal housing are individual penning or the acceptance of a reduction in average performance together with an increased spread in the range of performance. The solution chosen depends much on economics. Where a pig requires to be fed exact amounts, and where the amount may differ between pigs, then separate feeding stalls are obligatory. It is a short design step to mould an individual feeder into an individual pen.

Fundamental needs

Most pigs are farmed intensively, and intensivism is prone to generate animal welfare problems if the management is not expert. Criticism of intensive production methods is often framed in terms of criticism of the housing; particularly space allowances. It is impossible at the moment for animal well-being to be judged objectively, and it is difficult to avoid the logic that highly productive animals are unlikely to be ill-used. However, most of our farm animals are not so much 'highly productive' as 'productive enough'. Thus it can be the case that if animals were better used they would be even more productive. Whether that extra production is economic is another matter, and it is clear that when intensive systems do go wrong, the welfare of the animal is rapidly and seriously eroded. Codes of practice concerning space allowances, ventilation rates and so on can provide some general guidelines, but are of limited use in circumstances where particular house environments are so diverse, and both intensivism and extensivism can be equally associated with either high or low levels of welfare.

Baby pigs are born with no effective body energy reserves. They are also born wet. This makes them particularly prone to chilling, so they require a warm dry bed to lie on, warm air around and little or no air movement over the body surface. These problems are not shared by mother, so two micro-environments are needed in close proximity to one another. Indeed, high ambient temperatures tend to reduce the appetite of the lactating female, whose housing requirements do not extend much further than the need for individual feeding and ample water supply. Most of the paraphernalia of the maternity ward results from the need to keep sows confined on the one hand and the need to provide facilities for the baby pigs on the other. From the second week of age, the young pigs will also require feeding and watering devices.

The large number of house types available for newly weaned baby pigs is evidence of continuing failure to get this aspect of pig housing right. The needs of the pig are often subordinated to the needs for simple management. Thus a wire mesh floor deals with excreta more conveniently than a deep straw bed, but is less comfortable for the pig, increases behavioural and disease problems and requires more sophisticated environmental control.

Feeding arrangements tend to govern the design of houses for growing pigs. Floor-feeding saves space but requires solid floors. Troughs allow wet feeding by pipeline. Animals may lie on solid floors (with or without bedding), or on slatted or wire-mesh floors. They may excrete on to a strawed area, a solid concrete apron or through slots in the floor into a slurry chamber.

Growing pigs are nearly always – but not invariably – housed in groups of 8–40. In contrast, the combination of individual feeding and space saving has resulted in modern house designs confining adult breeding females in stalls or tethers, rather than in the more traditional loose-straw group-housing systems.

Examples of house types

About the only certain feature of pig housing is that design features are ephemeral. Figures 8.1 to 8.19 describe a number of typical house types. New, but not necessarily improved, house designs are coming forward all the time.

The environment

The nature of the environment bears heavily upon efficiency. Modern units try to control it.

Temperature

The definition of an adequate thermal environment is that which allows pigs to be comfortable: warm enough not to feel cold, cool enough not to suffer heat stress. Pigs interface with air, floor and each other, and it is upon these factors that it depends whether an animal feels cold. The *air* temperature and the rate of air movement affect the rapidity with which body heat is lost by convection and by evaporation to the surrounding air. The extent of *floor* insulation, and the floor type, affects the rate of transfer of body heat by conduction from the pig to the floor mass. The *density of stocking* has a positive relationship with both the temperature of the air in the house and the likelihood of one pig being able to lie

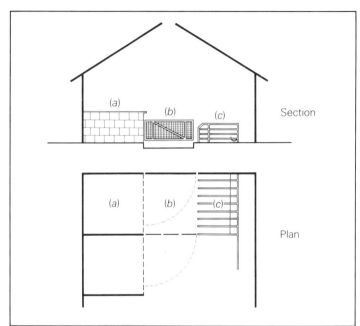

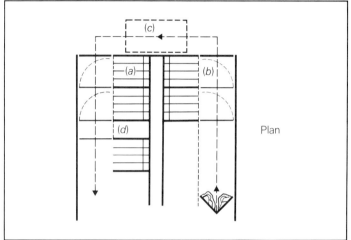

Figure 8.1 Loose housing for pregnant females. A deep straw bed is provided within a covered kennel (*a*). The yard (*b*) may be strawed and the excreta removed as a solid, or left unbedded and scraped off. A section is designed to hold 8 animals, each animal having an individual feeding stall (*c*). Water bowls or nipple drinkers may be placed in the yard. If females are moved here immediately after weaning an adjacent pen is given over to house the male

Figure 8.2 Covered kennels (*a*) provide both individual lying and feeding places for pregnant females held in batches of 4. Access to the feeders is through a hinged lid in the kennel roof. The kennel area may or may not be strawed. Animals excrete in a concrete yard (*b*) which is scraped automatically into a slurry tank (*c*). Water is best provided by nipple drinkers in the yard to avoid spillage over the lying area. A half-sized section (*d*) arranged in every third position is available for males, or females in need of special care

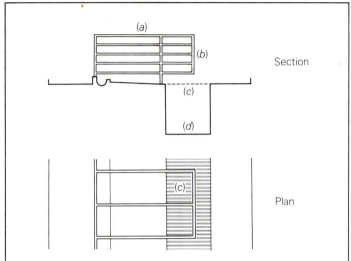

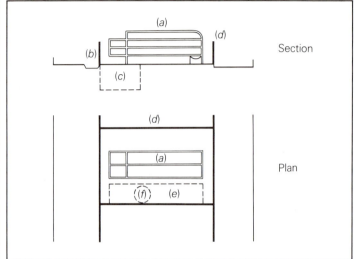

Figure 8.3 Stalls confine newly weaned and pregnant animals within a metal framework (*a*) which also serves as an individual feeder. Access to water is usually through a nipple located over the feed trough. If there is spillage, the pipeline may be opened for limited periods only. A gate to the rear (*b*) confines the animal, or a tether may be attached to the neck or girth of the pig and fastened to the floor or metalwork. Tethers allow greater freedom of movement, but both systems may cause skin sores. Excreta falls through slats (*c*) into a slurry chamber (*d*). Straw is not provided. Pens are located at the end of the house for males which are walked around non-pregnant females twice daily

Figure 8.4 Lactating females need to be separated from the special micro-environment needed by the baby pigs. This may be achieved with a bar partition or gate, or the female may be conveniently confined in a metal crate with feed and water trough incorporated (*a*). Straw – or better, wood shavings – may be given to the female, solid excreta removed by hand and urine run off into a drain (*b*). Alternatively, a slatted area and slurry chamber (*c*) may be fitted. The litter is confined by lightweight barrier walling (*d*). A special covered creep area or nest (*e*) can be provided; the essential feature being a source of extra warmth (*f*) and a high degree of floor insulation. Water and feed would be needed for the young pigs from 10 days of age

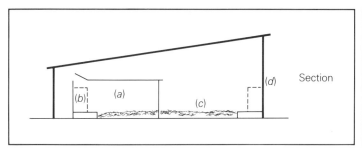

Figure 8.5 Traditional straw-yard suitable for young weaned pigs or growers (particularly to work weight at 50–70 kg). The kennel (a) is *deep* strawed and may also contain the ad-libitum feed hopper (b) set on a small plinth. Alternatively (or in addition) the feed hopper could be situated in the yard (c) with access from the front (d). Water is provided next to the feed troughs. Solid excreta and straw is removed from the yard. Pigs are housed in mobs of 20–40

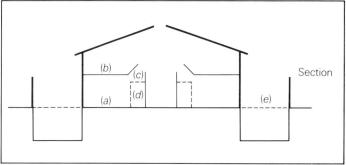

Figure 8.6 House for young pigs with solid floor lying area (a) which could be bedded with straw or shavings. A kennel roof (b) is provided over the lying area with a hinged panel (c) allowing access to ad-libitum feed hoppers (d). Pigs excrete on a slatted (or mesh) floor area which may be outside (e), as shown, or covered. Watering points are usually situated over the slats

alongside another and so provide a source of mutual warmth.

Pigs create heat as a natural part of body processes (half-a-dozen average-size growing pigs will throw off energy at the rate of about 1 kW h). Energy must actually be actively dispersed if their body temperature is to be kept normal at 39 °C. Heat stress results from a combination of factors conspiring to cut down the flow of body heat away from the pigs. For example, a high air temperature, a low rate of air movement, fully insulated floors and a high density of stocking. As a large part of the pig's heat is dissipated via the lungs by water vaporization, high humidity can also reduce the rate of body heat loss. Pigs react to heat stress by attempting to increase the rate of heat conduction by stretching out over the floor, by enhancing evaporative heat losses through wallowing and by increasing the rate of breathing. If the sun is acting as a source of radiant heat, shade is sought. If these stratagems fail, then, depending on the severity of the heat stress, pigs will eat less, grow more slowly, fail to rebreed, cease lactating and finally expire.

Cold causes pigs to lie on their bellies, to huddle close up to

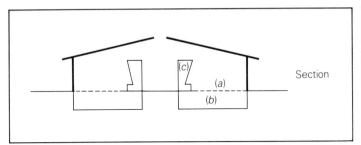

Figure 8.7 Intensive housing for young and growing pigs. The floor (a) is entirely of mesh, slotted metal or slats, with a slurry pit (b) below. No bedding can be provided. Ad-libitum feed hoppers (c) form the front of the pen and water troughs or nipple drinkers are provided at the rear. Pigs are usually housed in batches of 10–20

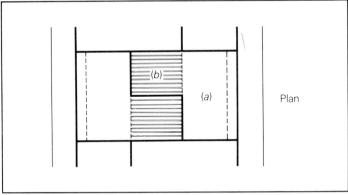

Figure 8.8 Solid floor pens (a) for growing pigs with slatted area (b) for excreta. Animals may be floor-fed automatically from hoppers held above, or troughs could be provided for pipeline feeding

one another, to shiver and to have an enhanced appetite. Because the sensation of cold is caused by an excessive rate of heat loss from the body, the pig reacts by increasing its rate of heat production. This is achieved by burning up food specifically to create heat. Whereas previously body heat was a waste end-product of normal life and growth, now heat creation is an end in itself, taking precedence over other productive functions. Cold animals therefore grow more slowly and convert feed less efficiently. They will also be thinner. Using food for heat production first diverts nutrients away from fat growth, and then away from lean growth. Where the food supply is inadequate to cope with the extra energy required to combat cold, fat stores – if they are available – are burnt up to create warmth.

Because 'cold' is defined as the temperature at which pigs divert nutrients from productive processes into body heating, there is no single defineable air temperature below which it could be said that a pig is cold. The amount of normal waste heat is relative to the current metabolic activity, which in turn is a function of the amount of nutrients supplied. As pigs are given more food, they not only grow faster, but they also

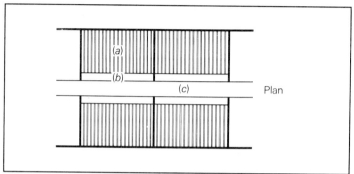

Figure 8.9 House for growing pigs with fully slatted floor (a) feed trough (b) and central feeding and access passage (c). Water is given with the feed, perhaps via a pipeline feeding system. Each pen may contain 8–20 pigs. Slurry is pumped out from below the slats

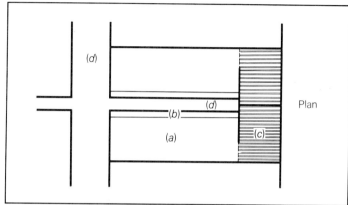

Figure 8.10 Solid floor pens (a) for growing pigs with trough allowing dry or wet feeding (b). Slatted area (c) provided for excretion with slurry chamber below. Access to the pens may be via the central (d) or slatted passages

become more resistant to cold (Fig. 8.20). As larger pigs are usually given more food, they are less demanding of a warm environment. This may often be misinterpreted to mean that because a pig is large it is also tolerant to cold, which is not so. Pregnant females may only be given a ration of 2 kg or so, and therefore throw off no more heat than a growing pig half their size. Cold tolerance relates to feed intake more than to body size. Loss of body condition and emaciation in breeding females may be associated with inadequate housing when the animal is pregnant.

Ventilation

As air moves faster through a pig house, the effective temperature will drop. An increase in the rate of airflow causes an increase in the rate at which convective heat is swept away from the pig's body, and also an increase in the rate at which evaporative losses occur (Fig. 8.21). The need for ventilation in pig houses is contentious, and rules about windows, fan size and fan speed can be misused.

Figure 8.11 House for recently weaned and pregnant females, with stalls and slats. The pens for the stock males are at the ends

Figure 8.12 Quarters for a lactating female and her young. Compared with the metal crate (Figs. 8.4 and 8.13) which has come to replace it, this house allows more freedom but is less convenient and more expensive

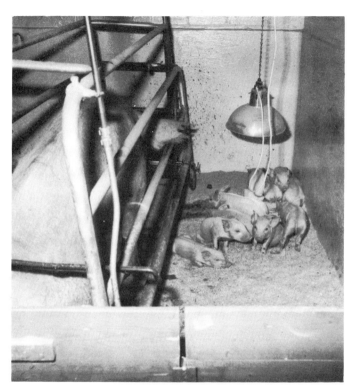

Figure 8.13 Conventional crate for lactating females

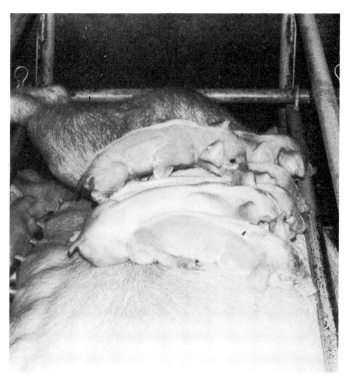

Figure 8.14 Slatted floors are the easiest way of removing excreta from a pen. The baby pigs make their own comment

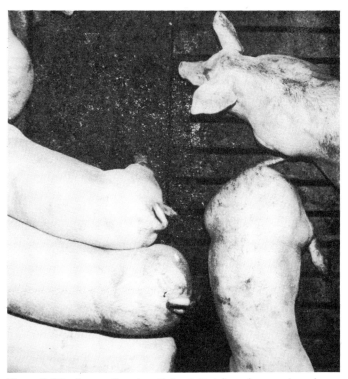

Figure 8.15   Deep straw weaner-pool with a protected nest to the rear

Figure 8.16   Conventional partially slatted floor for growing pigs

Figure 8.17 Automatic dispenser for floor feeding. The pens are slatted to the rear and hold 12 to 15 pigs each.

Figure 8.19 (a)

Figure 8.18 Pipeline system for wet feeding into troughs. The pigs lie in a bedded kennel area. To clean excreta from the pens, the doors are closed back and the passage scraped

Figure 8.19 (b)

Figure 8.19 (a) and (b) Contrasting extensive and intensive. Both systems have their problems. (a) shows an outdoor free-range system such as is still used as a 'break-crop' in some arable rotations.
(b) shows cages for baby pigs weaned at 10 days of age

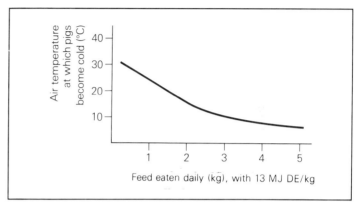

Figure 8.20 Influence of feed intake on the likelihood of a pig becoming cold

When there is a likelihood of heat stress, maximum fan capacity must be adequate to drive sufficient air through the house, both to reduce the temperature and to increase the possibilities for evaporative losses from the pigs. However, the chances of overventilation are greater than heat stress. The problem is usually one of keeping pigs warm rather than cooling them down. There is a certain illogicality in blowing warm air out of the house to replace it with cold, and the minimum ventilation rate is probably a function of the human threshold for dust, ammonia and noxious smells, rather than

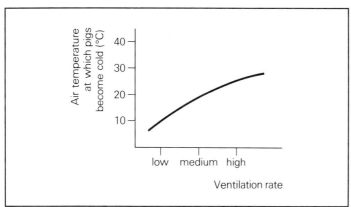

Figure 8.21 Influence of ventilation rate upon the likelihood of a pig becoming cold

Stocking density

The best way of warming air is with pigs. Adjusting the total weight of pigs (individual liveweight multiplied by number of pigs) per m^3 of airspace has the same effect as switching on or off electric heaters. More pigs give off more heat which allows a higher ventilation rate. Often pigs are cold because the airspace is too big, there is too high a proportion of small pigs or there is an insufficient number of pigs. On the other hand, smaller air spaces can be – indeed must be – ventilated faster; if properly done this can result in the environment being clean, dry, warm and free of smells simultaneously.

Floor type

Suspended floors, of wiremesh, slotted metal or slats, allow air movement around the whole body. The pig, although lying down, loses heat as if it was standing. The merits of suspended floors depend much upon the ventilation and air temperature characteristics of the house. If these are good, the pig will be more comfortable than lying on an uninsulated floor. But if not, it could be worse off with a suspended floor than with an insulated solid floor or a straw bed. Pigs lying on slats in over-ventilated, under-insulated houses are condemned to poor growth or high feed usage, or both. A deep straw bed, on the other hand, can offset a multitude of shortcomings in house design. Table 8.1 gives an indication of the extent of the effects of different floor types.

any need the pig might have for oxygen. Fresh air can come expensive on a cold winter's day.

The interaction between ventilation, water vapour in the air and building insulation is perhaps the most crucial factor governing the requisite rate of air movement. Pig's breath, warm and moisture-laden, builds up air humidity which, if not removed, will condense on the cooler wall and ceiling surfaces, so endangering the fabric. The less well insulated the house, the more probable the need to ventilate against condensation.

Table 8.1 House temperatures at which pigs given average ration allowances become cold (°C)

House conditions	Weight of pig (kg)			
	20	**40**	**60**	**80**
Ideal	21	17	14	10
Slatted lying area – medium ventilation rate	29	24	20	14
Insulated lying area – medium ventilation rate	26	21	18	13
Uninsulated – high ventilation rate – without straw	30	25	22	18
Uninsulated – medium ventilation rate – with deep straw	21	17	14	10

Notes: 1. Pregnant females become cold at about 15 °C if the floor is solid and there is a medium ventilation rate, higher if the house is draughty.

2. Newborn baby pigs can become cold at temperatures below 30 °C and require a supplementary source of heat.

The cost of cold

Estimates vary about exactly how much food a pig actually burns up when he is cold. But it is probably around 50 g of cereal equivalent per 100 kg of pig per °C of cold per day. Or, 2.5 kg daily for a pen of twenty 50 kg pigs in 5 °C of cold.

Smaller pigs are more susceptible, and here the cost cannot be counted in food alone. It is particularly difficult to sustain a temperature above 25 °C by natural means when the outside temperature is less than 15 °C. As the possibilities for increasing stocking density are limited, baby and young pigs require supplementary heating if they are to thrive and grow at all. It is usually considered economical to provide heaters from birth to at least 15 kg liveweight.

It is because of the different environmental requirements of different types of pigs that there *are* so many house types. Baby pigs require greater expenditure on their housing than young growers. It would be uneconomical to house baby pigs and 70 kg pigs together. Perhaps the most significant cost of cold is the price that must be paid for specialized housing, and a need for a large number of different houses of differing specification all on a single production unit; usually about five or six.

Diagnosis

A method of assessing housing adequacy is given in Fig. 8.22. The chart shows the central role played by the relationship between air temperature and ventilation rate. When all else fails there remains the choice of whether to increase the ration allowance or to put in supplementary heating. Heating should be a last resort, but if pigs are already being fed to appetite, which is usually the case at least until 30 kg, it may be the only option – albeit the most costly.

Standards

The final arbiter of successful housing design, and environmental adequacy is the performance achieved.

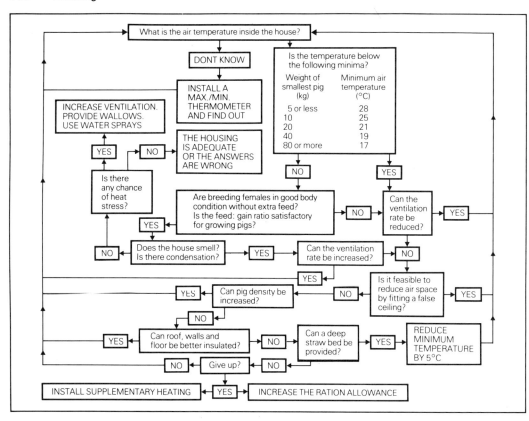

Figure 8.22 Diagnostic and action chart to assess housing

Table 8.2 Housing standards – quantitative

Space:	About 0.6–1.0 m²/100 kg of pig, total area for pigs in groups. Fully slatted or mesh floors, 0.5–0.8 m²/100 kg of pig. Adult pigs loose housed, 1.5–3.0 m²/100 kg of pig. Young weaned pigs loose housed, 1.0–2.0 m²/100 kg of pig Suckling adult with piglets, 4–7 m². Breeding males 7–10 m². Crates for adult females, 2,000 × 600 mm (approx.).
Air movement:	About 0.05 m/sec at pig height. 0.15–0.20 m/sec constitutes a noticeable draught.
Ventilation:	A minimum of around 0.3 m³/hr per kilo of pig weight. Maximum, about 2.0 m³/hr per kilo of pig. Average 1 m³/hr per kilo of pig housed.
Humidity:	70–80 per cent relative humidity.
Temperature:	In the region of: 26–30 °C for pigs of <10 kg 22–26 °C for pigs of 10–20 kg 18–22 °C for pigs of 20–50 kg 16–20 °C for pigs of 50–100 kg 16–20 °C for adult females.
Feeder space:	0.25 m/pig when 20–50 kg 0.30 m/pig when 50–100 kg 0.35 m/pig when >100 kg.
Excreta:	Average of about 5 litres/pig/day for a mixed breeding/growing unit.
House insulation:	To prevent more than 1 watt passing through the walls, ceiling or floor per m² per °C difference in temperature between one side and the other.

Table 8.3 Housing standards – qualitative

Space:	Enough lying area for all pigs to lie side by side, plus 25–30% for excretory functions. If pigs are likely to fight, enough space for the vanquished to escape. Animals restrained with tethers or in crates to have as much space as possible consistent with prevention of escape and attacks on neighbours. Excretion in the feeding area to be prevented. Body sores to be absent and the animal clean and dry (able to rest with body away from faeces and urine).
Ventilation:	Air movement over the surface of face and hands not readily detected. Air not fresh, but not stale nor unduly smelly either.
Humidity:	Not so dry and dusty as to cause respiratory distress, not so damp as to cause condensation.
Temperature:	Pigs < 10 kg: man comfortable with no clothes on. 10–20 kg: man comfortable in vest and pants. 20–50 kg: man comfortable in shirt and trousers. 50–100 kg: man comfortable in overalls but without coat. Adult females: man comfortable in overalls but without coat. Pigs lying huddled together are cold. Pigs lying apart are too hot. Pigs should lie side by side with flanks touching.
Feeder space:	To allow all pigs in pen to eat simultaneously and get their fair share.
Excreta:	Twenty per cent of the food eaten and all the water drunk.
Floor insulation:	Slatted or mesh floors to be free of through draughts. Solid floors to be dry, insulated and (unless

Table 8.3 Cont'd

	exceptionall well insulated) bedded. Wood shavings sufficient to form a solid mat of 10–20 mm between pig and floor. Straw to allow pigs to lie both on and in it.
General:	Pigs clean and dry, disease at low level, behavioural problems (cannibalism and so on) absent. House readily entered and worked in by staff and management.

Achieved results ought, logically, to override any standard recommendations. Nevertheless, Table 8.2 may be helpful as some guide to expected average standards. These are not given as a substitute for commonsense however, and the less authoritarian views expressed in Table 8.3 and Fig. 8.22 may be equally as useful.

Monitoring performance: records and control

9

Computerization – a look to the near future

It is not necessary to foresee a computer boom – it is already on. The extent of the service that modern electronics can give is wide ranging, and particularly significant to pig production which is both technologically advanced and in desperate need of a comprehensive number-crunching service.

Forecasting and response prediction

To keep on top of a rapidly changing situation, market intelligence is essential. This can take two forms: forecasts of the future economic climate, and prediction of how pigs will respond to a change in circumstances. Forecasting is fraught with difficulty, but the economists are gradually getting better at it. It is an important remit of the Government agencies and advisory services to provide forecasting intelligence. Response prediction is a quite separate discipline and depends upon complex computer programming which attempts to simulate the activities of live pigs as they grow and reproduce. These programs – or *models* – are dependent upon the current state of knowledge of pig biology, which although not particularly good, is now sufficient to furnish such programs as the Edinburgh Model Pig which yield the type of information contained in Chapter 6. The rapid proliferation of simulation models in the form of computer programs – *software* – will undoubtedly be a major feature of pig production in the 1980s.

It is even probable that most new knowledge coming forward will actually be packaged up in computer models, as the most convenient and rapid way of floating technical advances out to the industry and its advisers.

Sums

Computers can do complex sums quickly, they are particularly appropriate for financial calculations, profit and loss accounts, cash flows, wages and so on. This type of activity was the first commercial use to which computers were put and it is still this function that dominates our bank balances, if not our lives.

For the individual pig unit, financial calculators are a boon to time saving and error prevention. Already a range of pre-programmed software is available for this purpose, either for direct use on farm or indirectly through an advisory or consultancy agency. The way the mathematical abilities of the computer are accessed will depend upon the size of the farm unit and the type of computer involved; the bigger the farm and the cheaper the computer the more likely for the electronics – *the hardware* – to be physically present in the farm office – *in-house*.

Instructions

Computers can issue mechanical instructions by electronic control circuits. These can be generated automatically from the results of a previous computer calculation (perhaps a response prediction), or may more simply follow from a manual pre-setting.

A simulation model might suggest a particular nutrient specification for a diet. This would be offered to a least-cost diet formulation program which would come up with the required ratios of feed ingredients for the mill and mix unit to make up the diet. Diet formulation control (Fig. 1.1) could then be activated by computer instructions going direct to the flow controls. The size of the feed drops to growing pigs can also be put under computer control. The frequency of feeding can be programmed in, and the total daily ration allowance set. After telling the computer the size of the weekly increment in the ration scale, the electronics can be left to automatically provide the animals with their pre-set feed requirements right through to slaughter. As a bonus, complex ration scales become simple because the computer does it all. At the extreme, it would be feasible for a batch of growing pigs to have a different diet mix and a different ration every day, at no extra cost or effort on the part of the manager.

Records

Amongst the particular banes of any recording scheme are the problems of individual animal identification, the physical exertion of taking the record (particularly if it is for liveweight) and the hassle of retrieving the information from the mass of paper recording sheets when it is needed. Properly handled, computer science should be able to deal with this.

There is a reason to hope that soon a recording system could be built into the brick-work of a new unit. Animals would be given unique identification numbers readable by electronic eyes. Automatic weighing machines situated in passages between buildings would relay animal weight data back to the central computer, while information such as productivity and body condition of breeding females could be keyed in by an operative from the maternity quarters.

Carefully prepared software would not simply store the recorded information, but look at it and come back to the unit manager with guidance. Feed supplies into the buildings could be automatically monitored and feed conversion efficiencies and growth rates declared as weekly or monthly averages, or as a rolling annual mean. Breeding females failing to achieve required productivity standards would elicit a computer response to cull. As a weaned female passes through the weighing machine, the computer would assess her present weight and, after taking her condition into account, would calculate and instruct the requisite ration allowance for the pregnancy.

Much of recording is about counting days. How long is it since a female was weaned; when is another due to be moved out of her pregnancy quarters into the maternity ward; when will a batch of growing pigs be up to slaughter weight? As computers can count so well, the calendar presents no

problem to them and they can come up each day with management guidelines: female X has been weaned for 15 days, Y was mated almost three weeks ago, Z is 110 days pregnant, A has been lactating for three weeks, pigs in grower house B should be up to weight next week, and so forth.

The essential point about computerized records is that tedium is avoided by their collection, calculation and summarization being automatic.

Hardware alternatives

Computers can fill a building or a pocket. The large mainframe computers are powerful and can carry out complex tasks for a great number of users at high speed. Managers of pig units would have access to them either through a middle man, or would share time on their own account. If form-filling is involved at the interface between user and computer, some of the attractions are lost. A teletype console gives immediate access to the computer, either by a direct Post Office line or through a telephone handset.

Shared time is paid for as it is spent, so with heavy usage it becomes worthwhile to consider purchase of the hardware and putting it in-house. At the present time, advisory services, consultancy agencies and software specialists serving agriculture are tending to either time-share or use medium-sized in-house computers (Fig. 9.1). Individual farmers are unlikely to purchase these powerful machines in the first

Figure 9.1   Two medium-sized computers. (a), left, could be used by feed firms and agri-businesses; (b), facing page, is in use at the East of Scotland College of Agriculture.

(a) A medium-sized computer (centre) with peripherals. To the right is a console with a typewriter keyboard and a visual display unit (VDU). To the left is a line printer which allows a hard copy to be made of the output. The operator is putting a floppy disc into the central computer unit. Information and data are fed into the computer via discs and the keyboard.

(b) The Edinburgh computer has a large capacity for both calculations and storage; it also uses more substantial discs (one placed on top of the cabinet). This system is accessed from a number of terminals simultaneously. These are teletypes or VDUs with typewriter keyboards and are in other rooms of the same building, in the Farm offices and in the County Advisory Offices many miles distant

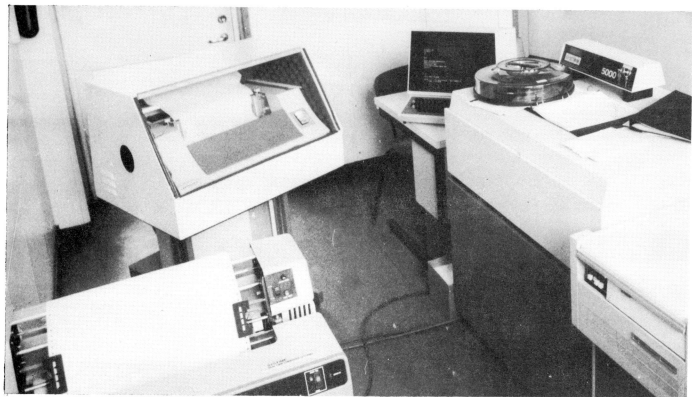

Figure 9.1 (b)

Figure 9.2 A desk-top computer with more limited scope, but of lower price

event, although co-operating groups and agri-businesses may well do so. The more probable first option for individual pig units is the desk-top type (Fig. 9.2). The smaller machines have the drawbacks of limited storage capacity and lack of calculating power. Often they are sold together with the specific items of software which fit them; for example, least-cost diet formulation and financial calculating programs. They are restricted in scope and it should not be assumed that newer and better software will necessarily fit into them. Many advisory agencies are considering the feasibility of developing links between a console in the farm office and the main-frame computer back at headquarters. This would allow the pig producer to have access to all the power and flexibility of the larger machines, but at low capital cost.

Records and recording

Records are for the purpose of improving the physical and financial performance of the unit. Contrary to general belief, they are not a favourite hobby of desk-bound managers, nor is their purpose to feed the office files with partly soiled paper. But it is regrettably true that in many cases the wrong records are kept, too many records are kept and too little is done with them. At the other extreme is the man who keeps no records – their being considered unnecessary while the unit is running smoothly. An acceptable view, only providing there remain no problems, and the manager is never put in the position of having to sort things out.

Records for selection

Breeders selecting for improved performance need information upon which to base that selection. As breeding is a long term and disciplined effort, the records need to be comprehensive. Breeding for faster growth and leaner carcasses can only be achieved if days to slaughter, weight and fat depth are carefully recorded for each individual animal in the various breeding lines. This sort of recording does not usually present any question of principle; one measures whatever criteria are appropriate. The problems that arise are more often of logistics and cost. In performance tests for males, an estimate of feed conversion efficiency might be thought necessary. This means at least individually feeding each pig to a set scale, and probably weighing back the refusals. Not a task to undertake lightly. Records of breeding performance, if comprehensively done, can also be consuming of time and effort.

Commercial records

It is not sensible to mount complex recording schemes for commercial herds. Apart from the costs involved, there is a major problem of individual identification of large numbers of animals. For most producers, recording should be kept to a minimum. Many of the problems of recording on commercial herds will, hopefully, be solved by computerization; but until that time comes the best possible must be done with paper records.

Two vital statistics can be identified for the commercial herd.

1. Pigs sold per breeding female. This encapsulates numbers born, re-breeding regularity and mortality. It is calculated simply from the knowledge of the number of pigs leaving the farm, per month or per year, divided by the total number of breeding females in the herd during the same period. A breeding female in this context includes maiden herd replacements from the moment they are selected at around 90 kilo.
2. Food used per pig sold. This is an overall feed conversion factor and measures the efficiency of utilization of the major raw material in the production process. It is calculated from the total tonnage of feed transferred into the unit, per month or per year, divided by the total number of pigs leaving the unit over the same time period.

Compendium efficiency measurements will indicate current performance levels and show changes in productivity, but cannot identify the causes of either of these. For diagnosis, closer examination needs to be made of specific aspects of performance. Pigs sold per breeding female may decrease owing

---

Pen record card   Pen number _____

Date at which pigs entered pen: _____
Weight of batch on entry (kg): _____
Number of pigs in batch: _____   Males: _____
                                    Females: _____
                                    Castrates: _____
                                    Total: _____

Diet given _____ from _____ to _____
           _____ from _____ to _____

Ration allowance:

Days	number of pigs	allowance per pig	allowance for pen
0–14			
14–28			
28–42			
42–56			
56–70			
70–84			
84–98			
98–112			
112–126			
126–140			

Sales from pen:

Date of sale	number of pigs sold	total liveweight

Figure 9.3 Record card for growing pigs between weaning and slaughter

Breeding female record card

Identification number: .... Card number: ....

Date of weaning: .... Parity number: ....

Weight[a]: .... Weight gain[b]: ....

Body condition score: ....

Male used .... Return date[e] ....

Date of first mating series[c]: ....

Date of second mating series[d]: ....

Ration allowance in pregnancy: ....

Expected date of parturition: .... Actual date: ....

Number of pigs born alive: .... Total weight of litter: ....

Number of pigs weaned: .... Total weight of litter: ....

Date of weaning[f]: ....

	weaning–conception	pregnancy	lactation	creep
Food consumption:			Date	

External parasite treatment: ....

Internal parasite treatment: ....

Erysipelas injection: ....

Pregnancy diagnosis: ....

Baby pig treatments – iron, teeth etc: ....

[a] If parity 1, animal is weighed at virgin mating.
[b] Weight gain is determined from the difference between weight at this weaning and weight at previous weaning (or virgin mating if parity 2).
[c] Give date of the first mating session of the two–three mating series.
[d] The second series often occurs three weeks or so after the first, if the first series should have proved ineffective. In practice constant vigilance over the assumed pregnancy is needed if females returning are to be noticed.
[e] Date of mating plus about 21 days.
[f] This date will also occur at the top of this individual's next card.

Figure 9.4 Record card for breeding females

Mating Register		
Date	Male used	Female mated

Food in/Pigs out register						
	Food delivered onto unit			Pigs transported off unit		
Date	Type	amount	source	Approx weight	number	destination

Figure 9.5 Record cards for matings and for amounts of food and pigs in and out of the Unit

to a reduction in the numbers born, a lengthening of the weaning to re-breeding interval, an increase in mortality, a high proportion of unmated herd replacements or a slowing down in growth rate. Inefficiencies of feed use might be in the growing or breeding herd, caused by wastage or cold, pig quality or feed quality.

A comprehensive recording system will allow instant identification of problems, but much of it may be under-used for much of the time. Devising paper records is simple: ensuring their effectiveness use is altogether more difficult. Three examples of paper recording cards are shown in Figs 9.3–9.5. The mating register remains in the mating area with the breeding males, the pen record card relates to each batch of growing pigs, while the card for breeding females would be kept in close proximity to the animal concerned, probably hanging above her pen.

Records can give direct information for action: movements from house to house, culling, dosing, weaning. Or more frequently they present a measurement of the performance level of the unit: number born to each female, bacon pigs sold, feed usage. Performance measurement is only useful if compared to some datum or target.

Targets

Average targets are listed in Table 9.1. These can be used to compare against the performance of an individual unit, but they may well be inappropriate:

1. The unit may be highly sophisticated and already have performances as good as the average target, even although above average performance is essential for the unit to be profitable.
2. The target could be so far in advance of the current performance of a unit that their possible attainment is not considered credible by the unit staff.

3. Targets may be inappropriate to one individual unit's circumstances; different ages at weaning, feed density, systems of production, house types and genetic merits of breeding stocks clearly require different targets. The best targets to be set are probably derived internally. Current performances can be identified and an achievable percentage added on. This will make for a gradual step-wise

Table 9.1 Examples of 'average' performance targets. Values relate to 100 breeding females

Litters per breeding female yearly	>2
Pigs born per litter	>11
Pigs weaned per litter	>9
month	>180
Birth weight of pigs (kg)	>1.2
Twenty-one day weight (kg)	5
Interval from weaning to mating (days)	3–5
Interval from weaning to conception (days)	<12
Number of parturitions monthly	18–21
Pig sales yearly	>1,900
monthly	>170
Breeding herd replacements, six-monthly	<25
Food used per breeding female yearly (t)	1.0–1.3
Total food usage by the breeding herd, monthly (t)	9–10
Food used by growing pigs yearly (t)	600
monthly (t)	50
Food used by each growing pig in one month (kg)	60
Age at 100 kg (days)	<180
Food conversion efficiency to 100 kg, medium density diet	<3.2
high density diet	<2.8

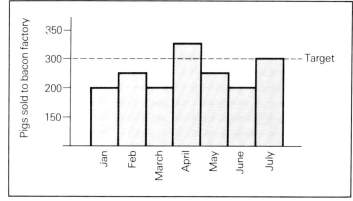

Figure 9.6 Histogram to measure pig sales. The chart is made up at the end of each month. Both actual performance and adequacy of performance is seen at a glance. Seasonal and time trends may also be distinguished. The same style of record can be used for number of pigs born, feed purchased, percentage of pigs in top grade, and many others. (Table 9.1 gives a list of 'average' performance targets)

progression upwards. Similarly, individual aspects of production efficiency could be singled out one by one for special attention. Current performance levels can be compared to targets by means of histograms, graphs or charts; examples are shown in Figs 9.6–9.8.

Management boards

One type is in the form of a solid rectangular board with days

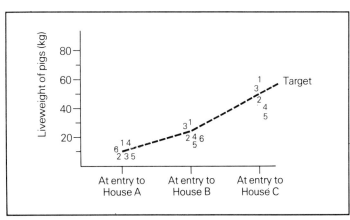

Figure 9.7   Graph to assess growth performance. Batches 1, 2 and 3 were on target, but 4 and 5 were poor. If 6 does not pick up in House B some investigation as to the causes of the reduced growth is indicated

of the year across the top and individual sows listed down the left hand side; there is a row to each breeding female in the herd. A cursor moves across the board as the calendar progresses. Time-based events such as mating are entered on the board, and the time of expected parturition and weaning are calculated and marked up. The board will then inform on the state of every pig on any day, and give warning of events due to befall her in the near future. A special mark can be placed against animals weaned and not proven to be pregnant so they can be given especially close attention. These boards tend to become tedious to arrange and be rather oversized if large numbers of sows are involved. An alternative, suitable for medium sized herds, operates on the principle of a large circular disc upon which a flag for each individual female is placed. The disc progresses in a clockwise direction, and as time goes by the animal in due course meets a mark indicating impending parturition, weaning, potential three-week return to oestrus, or whatever. These boards can offer day-to-day instructions and give warnings of some of the vital management factors in the breeding herd. But they require careful attention and upkeep.

Calculating housing requirements

Pen requirements depend upon the housing types selected and the performance of the pigs. The calculations are, however,

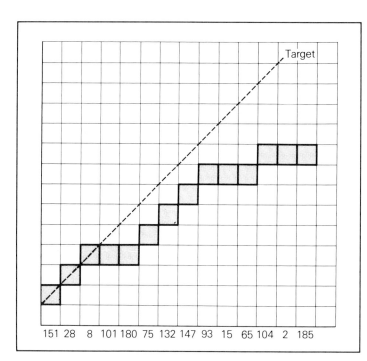

Figure 9.8 Chart to monitor effectiveness of maintaining standards. Standard successfully achieved – move up one square. Standard not achieved – move across one square. The chart is used here to monitor fertility. The standard is for pregnancy to be attained within 7 days of weaning. After 7 days have elapsed since weaning, each female's square is made up; if she is pregnant up one, if not across one. The increasing distance between the target line and the squares measures the extent of infertility. Pigs 151, 28 and 8 *were* mated within 7 days, 101 and 180 were *not*, 75 and 132 were, and so on. The chart shows standards to be slipping and a need for management intervention. The chart can be used for any performance measure which can be given a definitive target and a yes/no answer to the achievement of that target. For example, weight change of breeding females weaning to weaning (was it more than 15 kg?), age of pigs at slaughter (was it less than 160 days?), body condition of females, numbers of pigs born, pigs sold monthly etc.

simple and readily completed for each individual set of circumstances. Examples are laid out in Table 9.2. The major confusion factors in allocating housing are: bunching, variation in growth rates and reproductive performance *between* batches, variation in growth rates and reproductive performance *within* batches.

Bunches of breeding females reaching the end of pregnancy simultaneously cause havoc in the maternity ward. In some systems the pigs are grouped by design, so allowances are made in pen allocations. Variation is a natural phenomenon exacerbated by seasonal trends. Young growers might have been allowed a 50-day stay in a rearing house between weaning and 20 kg, but take 70 days before they are ready to

Weaning to 20 kg: 0.3 kg gain/day = 50 days[c]; (50 × 2.3)/365 = 0.33 or 33 per 100
  if each pen holds pigs from two litters, 33/2 = 16 pens
OR 50 days for 2,100 pigs = 105,000/365 = 288 pig places
  if each pen holds 18 pigs, 288/18 = 16 pens
OR 50 days = 365/50 = 7.3 batches per year
  if each batch is of 18 pigs, 2,100/(7.3 × 18) = 16 pens

20 kg to 100 kg: 0.6 kg gain/day = 133 days; (133 × 2.3)/365 = 0.84 or 84 per 100
  if each pen holds two litters, 84/2 = 42 pens
OR 133 days = 365/133 = 2.8 batches per year
  if each batch is of 18 pigs, 2,100/(2.8 × 18) = 42 pens

Weaning to 100 kg: 0.4 kg gain/day = 240 days[c]
  240 days for 2,100 pigs = 504,000/365 = 1,381 pig places
Weaning to 100 kg: 0.7 kg gain/day = 136 days[c]
  136 days for 2,100 pigs = 285,600/365 = 782 pig places
Weaning to 60 kg: 0.5 kg gain/day = 110 days[c]
  110 days for 2,100 pigs = 231,000/365 = 633 pig places

[a] 12–25 days, use values appropriate to circumstances
[b] 21–56 days, use values appropriate to circumstances
[c] 10 days safety margin
[d] For straw yards divide by number of pigs in each group, but allow ample spare capacity.
[e] With small safety margin

move on. Pigs ready to be weaned will have nowhere to go, while places in the finishing house remain unfilled. Management must do all possible to maintain a smooth flow of pigs through the unit, but the planners must allow pen spaces for peaks and troughs in production. Because of individual differences in performance, pigs in a batch will reach slaughter weight at different times. The dilemma for the

Table 9.2 Example of calculations for housing requirements (per 100 breeding females)

Allowing 115 days for pregnancy, 12 days for weaning to conception[a] and 28 days for lactation[b]. Assume nine pigs of 6–7 kg weaned/litter.

Number of litters/year:	365/(115 + 12 + 28) = 2.3
Number of pigs/year:	2.3 × 9 = 21 = 2,100 per 100
Stalls for pregnant females:	(115 × 2.3)/365 = 0.72 = 72 per 100
Maternity pens:	28 + 10[c] = (38 × 2.3)/365 = 0.24 = 24 per 100
Mating pens:	12 + 10[c] = (22 × 2.3)/365 = 0.14 = 14 per 100[d]
Boar pens and hospital quarters:	0.1 = 10 per 100 females
Pens for replacement breeding females:	Ten pig places or three group pens
Quarantine quarters – off unit:	Three pens

manager is to choose between emptying the pen all at once, accepting that the biggest pig might be 10 kg or more heavier than the smallest; or waiting for each pig to come up to exact weight, which might take a spread of three weeks. Sale contracts which allow pens to be emptied on an all-out basis have much to commend them.

Moving and mixing

Pigs grow, their pens do not. Because of this incompatibility, pens are either too empty or too full. Usually, pigs are moved from one pen size to another, or the number of pigs in the pen is adjusted. In most systems all three mechanisms operate. When first moved into a pen, pigs tend to be insufficiently densely stocked, while just prior to them being moved out they are too cramped. Pen sizes and group sizes are frequently changed. Each move tends to be associated with a growth check; the greater the change of circumstances, the greater the check. Pigs may be weaned into weaner-pools carrying mobs of 15 to 50. From these are drawn evenly matched batches of 10, 20 or 30 pigs for making up pens in the growing-finishing house. In some systems a few pigs may be dispatched for pork, thus easing the cramped quarters for the rest of the batch going on to 90 or 100 kg. Alternatively, pigs may remain in groups of 15 or so from weaning to slaughter, starting in small pens and being moved a couple or so times as they grow into pens of progressively greater dimension.

Pig mixing is encouraged by the need for matched batches of growing pigs to try to reduce pen emptying spread. It is particularly during early growth that pigs tend to be mixed with strangers. This causes tension, confusion and quarrelling, with consequent reduction in performance. Mixing is a managemental convenience which reduces efficiency and is best kept to a minimum.

Ill health

Most farms have some diseases most of the time. These must be countered either directly with routine medications, or better indirectly through providing an environment in which the presence of the disease does not materially affect pig performance. Other diseases come intermittently. When present, direct action against them is needed, often with the help of the veterinary surgeon. The best protection against invasion by these diseases is isolation of the unit from all foreigners; pigs and people. Both sorts of disease problem can act on the unit at clinical and subclinical levels. Clinical disease shows definite symptoms while the subclinical eats away at performance and efficiency levels without making itself obvious.

All ill health is a management failure: because the disease was allowed access into the unit in the first place, and because the environment allowed the organism to flourish. There can be benefits from the manager taking regular walks around the unit in the company of his animal production adviser and veterinary surgeon. In any event, immediate action and request for veterinary assistance should follow from groups of animals (rather than isolated individuals) going off their feed, breathing fast and heavily, vomiting, with diarrhoea, failing to breed, aborting, producing mummified foetuses, few live young at birth, many dead young at birth or themselves dying.

Foot-and-mouth, swine fever, swine vesicular disease, tuberculosis, brucellosis, Teschen diseases, anthrax and rabies are thankfully unusual in European herds. More common scourges of intensive pig units are Aujesky's disease, swine dysentery, transmissible gastro enteritis, vomiting and wasting disease, pneumonia, meningitis, salmonellosis and atrophic rhinitis. Swine dysentery and transmissible gastro enteritis are disastrous while active. But others, such as Aujesky's disease, pneumonia and atrophic rhinitis, can settle down on a unit to cause steady insidious performance losses. The rate of loss may range from being crippling to being of little practical significance. The means of ridding the unit absolutely of these pestilences is to clear out all stock and start again with specific pathogen-free animals derived by hysterectomy from minimal diseased pigs.

With help from the vet, many of the day-to-day diseases are dealt with within a normal management routine: internal parasites by injectable or oral dosing with anthelmintic before parturition (breeding females), or at weaning (young pigs); external parasites by regular skin dressing and care; diarrhoea in young pigs by additional antibiotics into the feed; baby pig anaemia by injection of a ferrous compound after birth; and erysipelas in adults by injection at weaning. Enteric infections of young growing pigs, while reducing performance, can be otherwise unremarkable and pass off; or occasionally can flare into a serious outbreak of intestinal disturbance causing death. Most outbreaks of digestive tract disturbances and diarrhoea can be controlled with high-level antibiotic dosing into the feed or directly down the pig's throat.

Two complex syndromes in breeding females, MMA (mastitis, metritis, agalactia) and SMEDI (still births, mummified foetuses, embryo death, infertility) are caused by a variety of organisms, including viruses, interacting with a variety of environmental circumstances. Overt systems of MMA in an individual female can respond to hormone and antibiotic injections, but the herd problem for both these syndromes can be frustratingly intransigent. After eliminating all possible known causes of reproductive failure, there is little else to do but attend to treatable symptoms and hope that the storm blows itself out.

Routine control

Management checks are best designed around the individual characteristics of the staff, buildings and production process.

On large units even the vital statistics are hard to keep up with. Whether or not a female is pregnant, when a litter is due for weaning, how often a male has been used, how old is a pen of growing pigs.

Maintaining control is simplified if information is incorporated into the physical structure of the unit; for example, a special area of the house set aside for females not yet pregnant, weaning always on the Friday after the pigs are three weeks old (or whatever), marking pens of weaners with expected moving-on date, and pens of growers with expected slaughter date.

As an approximation, a herd weaning at three weeks should have, on any given day, about 8 per cent of the herd's females in the mating area, 14 per cent suckling young and 78 per cent pregnant. For six-week weaning, equivalent proportions would be about 7 per cent, 25 per cent and 68 per cent.

Day-by-day tactical management requires a recording

Table 9.3 Checklist for parturient and lactating females

Observe	Consider
Days to parturition	Move to maternity ward after 105th day of pregnancy.
Cleanliness of quarters	Disinfect and rest for seven days between pigs.
Cleanliness of pig	Wash and scrub, treat with dressing against external parasites and mange. Treat with anthelmintic against internal worm parasites.
Temperature of house	Baby pigs need to be 5–10 °C warmer than their mother.
Welfare of pig	Correct pen arrangements.
Onset of parturition	Cost benefit of attending the birth.
Post-parturient problems	Lack of milk for baby pigs, high temperature (above 40 °C) fever in mother.
Availability of water	Automatic drinkers, regularly checked.
Days to weaning, reproductive performance of female	Record data in preparation for decisions to move, cull or remate.
Routine injections	Erysipelas (not while pregnant).

Table 9.4 Checklist for sucking baby pigs

Observe	Consider
Completion of routine tasks within three days of birth	Administration of iron compound. Clipping eye teeth. Ear mark (to record at least the week of birth).
Completion of routine tasks within 10 days of birth	Castrate if necessary. Provide highly palatable creep feed diet for baby pigs. Keep fresh by twice daily provision, and attention to cleanliness of trough. Check water supply to baby pigs.
Signs of diarrhoea	Antibiotic dose. Improving standards of cleanliness.
Environment	Bedding, temperature, lack of draughts.
Growth performance	Reductions in suckled milk supply can be offset by increased intake of dry creep feed.

system together with close observation of the current physical status of the animals on the Unit. Tables 9.3–9.8 convey something of the spirit of a Unit Manager's routine mental checklist.

Table 9.5 Checklist for breeding males and females awaiting mating

Observe	Consider
Males	
Size	Ration allowance.
	Age.
Willingness to mate	Age.
	Health.
	Ration.
	Frequency of use.
Numbers born	Frequency of use, etc.
	Use two males on same female as insurance policy.
Disposition	If aggressive, check the activities of the staff. Check water supply.
Females	
Days since weaning	Which at fault? Pig, ration, diet, disease or stockman.
	Culling policy.
	Housing type.
Body condition	Ration in lactation.

Table 9.6 Checklist for pregnant females

Observe	Consider
Body condition and weight-change weaning to weaning	Pregnancy ration.
Confirmation of pregnancy	Check with males 18–25 days after 1st matings.
	Check with pregnancy diagnosis equipment.
House temperature	Use of max : min thermometer.
	Draughts.
	Behaviour of animals.
Welfare and restlessness	Environment.
	Concentration of diet; provision of roughage.
Cleanliness, excreta on hind quarters	Penning arrangements.
Physical abrasions	Penning arrangements.
	Group antagonisms.
Time since mating	Movement to maternity quarters.
Feed and water supply	Automation.
Slow eating rate	Disease.

Table 9.7   Checklist for weaned pigs

Observe	Consider
Adequacy of environment	Provision of plenty of space, but not such the animals are in any way cold.
	Possibilities of deep, clean straw.
	The very best intensive housing.
Diarrhoea	Medication of feed.
Feed provision	Supply of more feeding space, and fresher more palatable feed.
Growth rates	Temperatures, disease, welfare, feeding arrangements, size of group, stocking density.
External and internal parasites	Dress and dose.

Table 9.8   Checklist for growing pigs

Observe	Consider
Cannibalism	Environment, bedding material, diet.
Feeding arrangements	Adequate space.
Slow growth, variable growth rates within pens	Disease.
	Too frequent mixing and moving.
	Ration or diet.
	Stocking density.
	Environment.
	Disease.
Slow eating rate	Checkweigh.
Weight for slaughter	

Tailpiece

More meat is being eaten, and of this more is from pigs. But if pig production is to keep its place in the agricultural economy, efficiency must continue to improve.

Pigs are no longer fat animals, so reduced fat depth is not a priority breeding objective. Attention will now be toward pigs with faster rates of lean growth. Reproductive performance has not improved much over recent years; the UK average number of pigs weaned per sow per year is still less than 18, despite 22 being a readily achievable target. Perhaps both farmer and geneticists should look harder at the breeding herd.

Marketing would be simplified from the producer's end if the product were apportioned to the various outlets after the animals left the unit, and not before. This would allow pens and houses to be emptied completely in a single day, rather than piecemeal.

The state of knowledge of growing pig nutrition is comparatively well on, and current technology is directed to optimization of nutrient supply and strategic feeding. This is where the computer comes into its own. Feeding breeding females is quite another matter, and there is still a serious shortfall in objective information upon which pig producers can make the right nutritional decisions.

Perhaps the most concern should be with the care of mother and young after weaning. Reproductive inefficiency is most frequently a result of failure to rapidly rebreed, while the young growing pig performs at far below his potential.

The welfare of the pigs is bound to receive an increasing share of attention. Not only because it is morally right, but also because profitability will come to depend upon it. Currently, pigs are not kept particularly well, merely well enough.

It is unrealistic to believe that in future an intensive unit can depend for its survival upon intuitive stockmanship. There must be an increasing move toward better and more complete automatic computerized systems for the care of pigs. This need only be an alarming prospect if the job is badly done. In the hands of skilled technologists it will enable better conditions for the animals to live in and an improvement in the animal's lot. If well served by their keepers, pigs will return the compliment by efficient and profitable production of high quality human food.

Index